MODERN PYROMETRY

By

CHARLES H. CAMPBELL, M.S., Met. Eng.

Research Engineer, American Steel and Wire Co.,
Research Laboratory

1951

CHEMICAL PUBLISHING CO., Inc.

212 Fifth Avenue   New York, N. Y.

Copyright
1951
CHEMICAL PUBLISHING CO., INC.
NEW YORK N. Y.

FOREWORD

This book was developed from a lecture prepared for the Cleveland Chapter of the American Society for Metals, that was given in the 1947 Spring Educational Series.

The lecture was designed to illustrate the relatively common as well as the new techniques in pyrometric practice and equipment in order to help those men who returned to industry recently as well as to acquaint newcomers in metallurgy with the principles of and new developments in pyrometry. No criticism of any particular pyrometric system was intended and, although all types and makes are not discussed, the selected group was deemed sufficient to give a fairly accurate picture of the available methods of temperature measurement and control.

I wish to acknowledge the assistance received, particularly from Gordon Spare who did the photographic work. My thanks are also due to the various authorities who have granted permission to publish this book.

CONTENTS

Chapter 1

TYPES AND USES OF THERMOCOUPLES, LEAD WIRE, AND PROTECTION TUBES

The improvements in the heat treatment of steel within the past few years require close temperature control on a large production basis. This is especially true of the isothermal spheroidization of high-carbon steel, which converts the iron carbide into spheroids or balls within the soft ferrite matrix. Such a structure yields a soft steel. Uniform temperatures must be maintained throughout the entire charge in this type of box-annealing cycle to produce a strip that will satisfactorily withstand a severe flat bend. If any portion of the charge overshoots the required temperature by 20 or 30°F., the resulting steel microstructure will contain some small brittle spots which produce breakage during a bending formation. Proper temperature control is also a prerequisite in the modern continuous annealing and patenting lines for the development of special properties in steels. The patenting treatment is used for wire to produce a microstructure of very fine pearlite. This structure has maximum ductility for wire drawing. For uniform nondirectional properties in deep drawing steels, for maximum decarburization and strain relief in electrical steels, for uniform patented structures in spring wire, for best physical properties in quenched and tempered products, close control of both temperature and time is required.

Concurrent with the improvements in metallurgical processes has been the development of more accurate means of indicating and controlling temperatures. Various types of instruments now being used will be described with regard to their mode of operation to illustrate to those engaged in pyrometry the new methods and techniques of securing accurate control.

Thermoelectric Effects

Before describing some of the recent developments in the technique of accurately measuring and controlling temperatures in the metallurgical field, it might be justified to briefly recount some of the fundamental concepts upon which pyrometry is based.

The basic discovery of pyrometry was made by Seebeck in 1821. He noted that if the ends of a copper and an iron wire were fused together and one of the junctions was heated, a current flowed from the copper to the iron wire at the hot end and from the iron to copper at the cold end. Peltier later observed a thermoelectric effect which was the reverse of the Seebeck discovery. This effect is observed when an electromotive force is applied to two dissimilar metals connected together. When copper and iron wires are used, a current flow from copper to iron produces a cooling of that junction while at the iron-copper junction, heating occurs. This heating effect at the iron-copper junction is distinct from that produced by the resistance of the wire. The extent of heating and cooling effect is dependent on the metals used and on the amount of current. A supplement to the Seebeck effect is that discovered by Thompson. This effect is illustrated by heating the end of a uniform copper wire. An electromotive force is developed between the hot and cold ends of the wire the magnitude of which depends on the metal and on the differences in temperature at the wire ends.

Considering these thermoelectric effects, a thermocouple may be defined as a pair of dissimilar conductors joined so as to produce an electromotive force when the junctions are at different temperatures. If one junction is maintained at room temperature or at the temperature of melting ice, the temperature of the other junction can be determined by measuring the electromotive force developed in the circuit. The electromotive force values developed by thermocouples are small, usually a few thousandths of a volt.

A simple circuit illustrating the thermoelectric principle is shown in Figure 1. Two dissimilar materials A and B are connected together at hot junction T_1 and cold (reference) junction T_2. A millivoltmeter placed in the circuit measures the electromotive force produced when there is a temperature difference between T_1 and T_2. The current flow must be very small in order to secure maximum voltage in the pyrometer circuit. When the null-balance potentiometer method is used, the current flow is zero at the point of balance. Therefore, the voltage measured is a maximum. This gen-

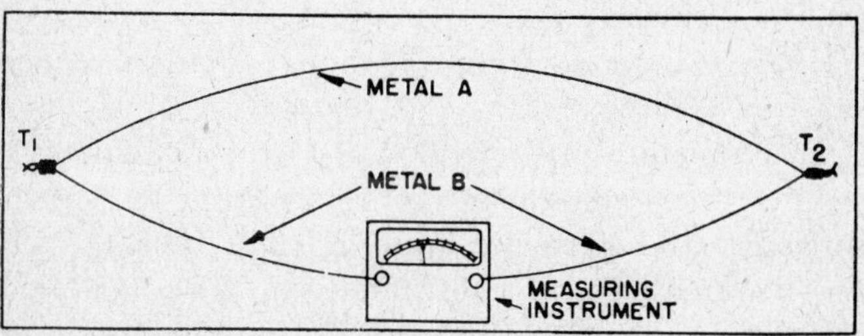

FIGURE 1. *The Fundamental Thermoelectric Pyrometer Circuit*

erated millivoltage is dependent on the kind of material used in the circuit and the temperature differential. Neither the diameter of the thermocouple wires nor variations in the temperature gradient along the wires have any effect on the output of a uniform couple. For example, a Chromel-Alumel couple will generate a definite amount of electromotive force for a given temperature differential whether the wire size is 12 or 20 gage. The heavier wire withstands rough handling and severe oxidation to a greater degree and is used on equipment not requiring rapid temperature response. However, if the wires are chemically or mechanically (degree of cold work) heterogenous over a given thermal gradient, the output of the couple will be distorted because of the establishment of secondary hot and cold junctions. The normal arrangement for a thermocouple circuit is to have the cold junction of the thermocouple connected to the compensated lead wires which go to the recording instrument. The cold-junction compensation is made at the instrument.

Requirements and Types of Thermocouples

Although any two dissimilar conductors may be used for a thermocouple, there are certain requirements which must be met if the couple is to be used commercially. There are five factors which should be considered in the selection of thermocouple materials.

    1. Capacity for resisting corrosion, oxidation, reduction and melting.

Heat will naturally accelerate any of these reactions. We know that the requirements of temperature measurement in an ingot-soaking pit are quite different from those of a tempering furnace or of a cleaning bath. If the couple does not have the ability to withstand the respective deteriorating effects, premature failure will occur at the bead or at a localized hot spot along the wire. This type of failure can be easily detected since the thermocouple wire is broken and the circuit will be open. However, in some cases, the couple may become contaminated with metal vapors or various furnace atmospheres and still continue to operate for a considerable time. This contamination will change the calibration of the thermocouple thus giving inaccurate temperature readings. Platinum couples are prone to such contamination. The use of various metallic and nonmetallic protection tubes has lightened the burden on the thermo-couple wires in this respect.

2. Development of a comparatively large electromotive force.

It is quite a task to measure low voltages accurately, therefore, a thermocouple with a high millivolt output would be preferable. According to thermoelectric tables, a couple composed of germanium and silicon would give a relatively high output, but it would be disadvantageous when other factors, such as corrosion resistance, strength and cost, were considered. The best couple will have the highest electromotive force at the desired temperature and also suffi-cient corrosion and oxidation resistance for the particular application.

3. Such temperature-millivoltage relationship that the millivoltage increases fairly uniformly with increas-ing temperature.

The importance of this point can be seen readily. Some conductors develop a maximum electromotive force at a certain temperature, but as the temperature continues to rise, the output decreases. In this case, the couple would indicate two temperatures at which the same electromotive force is generated and, therefore, trouble would be en-countered in controlling or recording. If the slope of the temperature-thermoelectric potential curve for a material changes rapidly it is difficult to secure uniform accuracy in controlling temperature. A large voltage change over a small temperature range would result in close temperature control.

4. Cost.

It is obvious that noble metal couples, in spite of their high resistance to corrosion and oxidation, are not suitable for relatively low temperature work, because of their high cost. But for high temperature work, the platinum couple is the only solution.

5. Reproducibility.

The thermoelectric properties of most materials are quite

difficult to control. It requires a close control of composition and physical working to reproduce a given millivolt output. This is particularly true of iron wire. They must meet definite thermoelectric specifications, and small variations in the residual aluminum, copper, or carbon content are sufficient cause for rejection. The effect of a slight composition variation on the millivoltage output of iron wire is illustrated by the following two samples.

Carbon	Manganese	Phosphorus	Sulfur	Silicon	Copper	Millivoltage
0.03	0.17	0.008	0.021	0.009	Trace	$+0.356$
0.03	0.08	0.006	0.022	0.002	0.10	-0.190

The millivoltage values show the plus and minus deviations from the standard iron wire at 1500°F. The difference in millivoltage characteristics of these samples was due to the variation in copper content. Thus without good control each couple would first have to be given a complete calibration, and the entire controlled process would be based on this particular couple. Since there would be a wide variation between couples, only the millivolt scale could be used. The temperature conversion would then be made from the calibration curve.

The four thermocouples listed in Table 1 have been recognized as standards for normal industrial uses.

High quality and uniformity of both thermocouple and lead wires are of prime importance for industrial application. At present, standard millivolt versus temperature curves are available for platinum-platinum rhodium and Chromel - Alumel thermocouples. The iron - constantan couples have not been standardized as yet because of the difficulty in getting pure, uniform iron and constantan. However, operating curves have been established by the manufacturers to which they adhere quite closely, the allowable error for iron-constantan being ± 0.5% from 1000 to 1600°F. The Chromel-Alumel guarantee is ± 0.75% from

TABLE 1

Standard Types and Limitations of Thermocouples

Type	Kind of Wire	Recommended Temperature Limits and Approximate Maximum Millivoltage	Atmospheric Conditions for Which the Thermo-couple Is Best Suited
Iron-Constantan	Iron—Positive Constantan—Negative	1600°F.—50 mv.	Reducing or Neutral
Chromel-Alumel	Chromel—Positive Alumel—Negative	2000°F. Continuous Duty—45 mv. 2300°F. Intermittent Duty—51 mv.	Oxidizng or Neutral
Platinum-Rhodium	Pure Platinum—Negative 10% Rhodium 90% Platinum } —Positive	2700°F. Continuous Duty—15 mv. 2850°F. Intermittent Duty—16 mv.	Requires Protection from All Atmospheres; Do Not Use Bare
Platinum-Rhodium	Pure Platinum—Negative 13% Rhodium 87% Platinum } —Positive	2700°F. Continuous Duty—17 mv. 2850°F. Intermittent Duty—18.4 mv.	Requires Protection from All Atmospheres

660 to 2000°F., while that of the platinum couple is ± 7° from 0 to 2400°F. Figure 2 shows the millivolt versus temperature curves for the various thermocouple wires, using pure platinum as the base standard. As indicated by the dotted lines, these calibrations are actually curves. The data for this type of curve are secured by welding the unknown wire to the pure platinum wire and observing the millivolt output at various standard temperatures. These standard temperatures are obtained through the use of

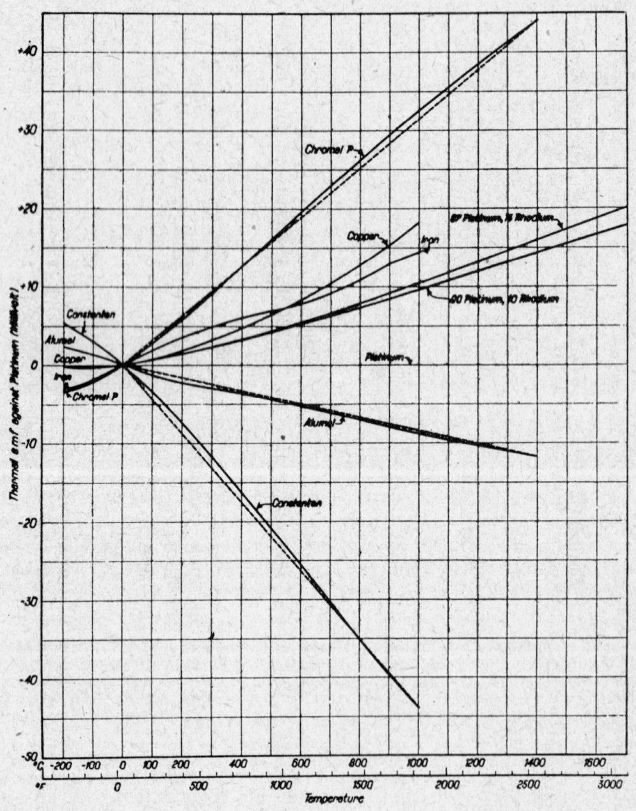

FIGURE 2. *Thermal Electromotive-Force Curves, against Standard Platinum, of the Principal Industrial Thermocouple Materials*

various pure metals having accurately known melting points. The hot junction of the standard platinum wire and the unknown wire is placed in a crucible containing a pure metal having a sharp melting point. This metal, after being melted, is permitted to cool slowly. At the temperature of solidification, the millivolt output of the unknown couple is read from a precision potentiometer in the circuit. Care must be exercised to maintain a constant cold-junction temperature. This can be secured by cooling to a constant 32°F. with an ice-water jacket. The millivolt output is determined for a series of temperatures, using different metals and the values are plotted against temperatures. The difference between such millivolt-temperature curves of any two metals indicates the millivolt output of an equivalent thermocouple within the same temperature range.

In order to secure the maximum uniform output from a thermocouple, there must be good contact between the two wires at the hot junction. For certain applications, the two wires of the couple are pointed and kept separated at the hot junction. Contact is made by holding both prongs against the heated metallic object or submerging them in a molten metal bath. Care must be taken to insure penetration below the scale or slag surface. This method of securing the hot junction of a thermocouple, using the heated object as the connecting link for the two prongs, yields an average value for the temperature between the two prongs. With this type of hot junction, an additional thermoelectric effect occurs between both prongs and the material contacted. However, its effect at the positive prong of the thermocouple is canceled by the opposing millivoltage at the negative prong.

For most applications, however, the two wires are welded to form a small bead at the hot junction. First, the wires are twisted together at one end (about two turns for stiffness) then fused with an acetylene torch or electric arc. Where a large number of thermocouples are being made, a device of Figure 3 is quite convenient. The positive terminal of a direct-current, self-excited generator is attached to the thermocouple wire, while a carbon block, located at the top of the T tube, is negative. The twisted end of the couple is inserted in the tube and an arc is struck with the carbon plug. By adjusting the voltage and arcing time, the proper degree of fusion can be secured. The entire unit is flushed with a protective atmosphere, thus yielding a good weld, free from oxides or slag.

Depending on the requirements, a thermocouple unit can

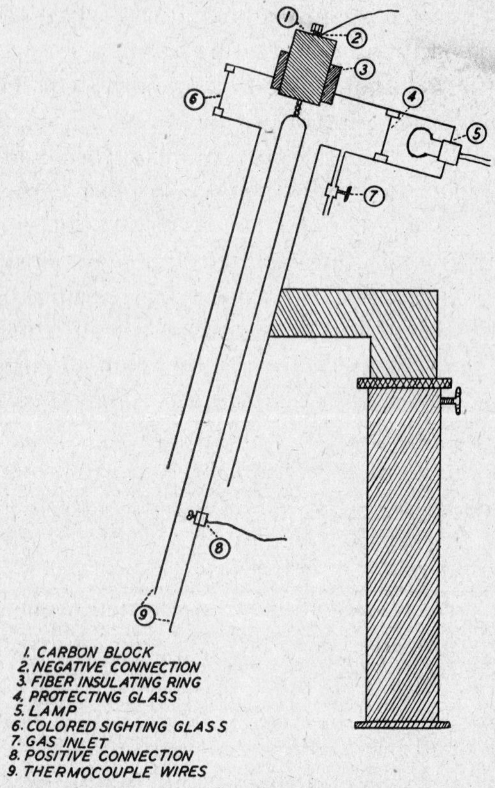

FIGURE 3. *Sketch of the Equipment Used to Arc-Weld Thermocouples under Controlled Atmosphere*

be assembled in various ways as shown in Figure 4. The illustrated couples have an increasing amount of protection from top to bottom of the figure. The bare thermocouple at the top is rarely used. For relatively low temperatures (below 1200°F.), where severe corrosion conditions are not encountered, the next type of couple, being electrically insulated and protected with asbestos, is satisfactory. The third couple with ceramic insulators is used for higher temperatures, e.g., to check the temperature of the charge inside

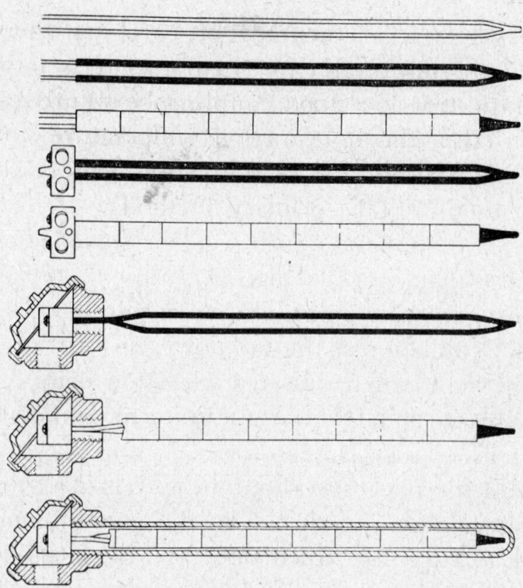

FIGURE 4.   *Thermocouple Insulation and Protection*

a box-annealing furnace. It is flexible enough to be used under the box annealing cover or to be pulled through a continuous furnace to obtain the data for a temperature-distribution curve. The terminal block and head connections insure good electrical contact and enable the couples to be replaced quite easily. The terminal connections between the couple and lead wire should be of suitable material, otherwise an additional, nonuniform electromotive force will be generated which will upset the standard calibration of the couple. The negative side of the thermocouple must be connected to the negative wire of properly matched leads. The bottom couple has the protection tube inserted in the terminal head and illustrates a completed unit for most medium- and high-temperature applications. A similar unit is used for liquid baths.

Thermocouple Calibration

There are several methods of checking thermocouple wires and thermocouples for accuracy. It is important in all calibrations that a good coupling be secured at the hot junction. Also, the hot-junction temperature must be accurately known and the cold-junction temperature maintained constant. The primary method has already been described in conjunction with securing standard calibration curves, i.e., that of checking the millivolt output of the couple at the melting points of various pure metals. Another technique is to connect the unknown wire to the bead of a Bureau of Standards calibrated platinum couple. This assembly is placed in a furnace at a uniform temperature. The furnace temperature is determined by the platinum couple after which the platinum rhodium wire is disconnected and the unknown wire is attached to the potentiometer. This produces a couple of which one wire is platinum and the other is the unknown wire. The resulting millivolt output can be checked with the standard curves. The connection of the unknown wire should be a weld at the hot junction of the standard couple or to a short length of platinum wire previously welded to this junction. The additional short piece of platinum lengthens the life of the standard couple.

If we connect a standardized wire, checked by one of the described methods, with an unknown wire of similar composition and we place this junction in a furnace of uniform temperature, the difference in millivolt output will indicate the extent of deviation of the unknown from the standard. If both wires are identical no electromotive force will be generated. This test is widely used for checking coils of thermocouple wire prior to actual thermocouple assembly.

When it is desired to check new coils of iron and constantan wire, thermocouples are made and tested to determine their matching characteristics. In this manner, a coil of iron wire which has a positive deviation from the specification could be matched with constantan wire having a similar error. Thus the resulting couple will yield the proper millivolt-temperature relationship for standard iron-constantan couples. These couples are usually made and tested in a group of eight or ten. They are compared with a couple of similar composition that has been calibrated against the standard platinum couple. All of the hot junctions are placed in a furnace uniformly heated at approximately 1300°F. for 2 hours and the millivolt output of each checked with a potentiometer. The positive or negative corrections can be observed and the wire properly tagged. The calibrating furnace can be set close to that temperature which is used in actual production operation. If this method of testing is used, all thermocouples must have their cold junctions at the same temperature. In routine checking, it is sufficient to have all cold junctions close together at a distance of 8 to 10 feet from the small testing furnace. Care should be taken that there is no gas or air circulating around the thermocouple junctions in the furnace otherwise erratic results will be obtained.

Types and Insulation of Lead Wire

In the majority of cases, the controlling or recording instrument is a considerable distance from the thermocouple. Therefore, it is general practice to connect the two by means of lead wires. This apparatus must give the same results as if one long continuous couple were used. The cold junction of a couple used in a furnace will be relatively warm; therefore, the lead wires must have thermoelectric properties equivalent to those of the thermocouple so as to produce no change in millivolt output of the couple as received at the control instrument. The leads transfer the cold junction from the relatively high temperatures around the furnace to the lower and more stable temperature at the controlling instrument. Here the cold-junction corrections can be made automatically or manually.

For the less expensive couples, such as iron-constantan and copper-constantan, the same compositions are used for the lead wires as for the thermocouples themselves. At temperatures below 250°F., the millivolt output of Chromel-Alumel is about the same as that of copper-constantan. Therefore, to achieve economy, copper-constantan can be used as lead wire when the cold junction of the more expensive Chromel-Alumel couple is below 250°F. When the cold junction temperature exceeds 250°F. the lead wires should also be Chromel-Alumel, thus eliminating any errors that would be produced by the junction of dissimilar metals. Special alloy leads have been developed for use with the platinum-platinum rhodium couples.

Since the maximum output of a thermocouple seldom exceeds 0.05 volts, the lead wire must have uniform quality and good electrical insulation to prevent leakage and galvanic currents. Three types of insulation are used, depending upon the specific application.

1. Polyvinyl thermoplastic is used for applications in which the temperature of the lead wire does not exceed 140°F. It is also resistant to deterioration and moisture.

2. Felted, wax-impregnated asbestos withstands excessive moisture and high-temperature flame.

3. Rubber. Flexibility and moisture resistance are the two important advantages of this insulation. However, the lead-wire temperature should not exceed 110°F.

The length of the lead wire may also have an effect on the accuracy of the recorded temperature, depending on the type of indicating instrument used. With the millivolt-meter-type indicator, the external resistance (that of the thermocouple and lead wire) should be as low as possible. The millivoltmeter is composed of a simple moving-coil galvanometer in a permanent magnetic field. A calibrated high-resistance coil is connected in series to the movable coil. Thus, the movement of this instrument is produced by the flow of a small current through the movable coil. If the resistance of the thermocouple and lead wires is excessively high, the voltage generated by the thermocouple at a given temperature will produce a reduced amount of current. Therefore, when the millivoltmeter is connected across the lead wires, the effective current going through the movable coil will be further reduced. The net result is a low voltage reading which may be corrected by decreasing the amount of calibrated resistance in the millivoltmeter instrument. This tends to decrease the accuracy and response of the millivoltmeter indicator. A 14-gage (Brown and Sharpe) lead wire is used with this type of instrument. The length is kept to a minimum to insure low electrical resistance. With the galvanometer-type potentiometer, this condition is not so crictical and a 16-gage wire can be used. In this case, the degree of accuracy is not dependent on the external resistance of the circuit since no current is required when the galvanometer is exactly balanced. However, with a long lead wire, the increased resistance decreases the current flowing from the thermocouple and lead wires to the unbalanced potentiometer. The amount of current drawn from the thermocouple circuit is decreased to zero by the application of a counter potential during the balancing of the potentiometer. At the point of balance, no current is flowing, so that the increased external resistance has no effect on the accuracy of the instrument. However, the galvanometer tends to become sluggish and longer time will be required for balancing. Where extreme lengths of lead wire are necessary, the fully electronic instrument is accurate and allows rapid balancing. In this case, the amount of current drawn from the thermocouple circuit is negligible and thus the effect of external resistance is eliminated. Leads of 20-gage wire or smaller may be used with no harmful effects.

Protection Tubes

In general, platinum-platinum rhodium couples have a high resistance to oxidation and corrosion, but these noble metal couples may appear to be in good condition and still be considerably off calibration. Platinum couples are contaminated by hydrogen, carbon and many metallic vapors. In the case of Chromel-Alumel couples, best results are obtained in an oxidizing atmosphere in which a thin oxide layer is formed that protects the metal against rapid scaling. However, when used in reducing atmospheres, difficulties arise from the preferential reduction of nickel oxide on the wire surface. If the couple has been previously oxidized, the reducing atmosphere will not affect the chromium and aluminum oxides, but will reduce the nickel oxide. This surface film of reduced nickel is a source of trouble, because the temperature gradient along the couple will produce a thermoelectric effect between the reduced nickel and the base metal. This parasitic thermoelectric effect generates a series of countercurrents which will distort the final measured output of the circuit. Iron-constantan couples should be used with a reducing atmosphere to prevent rapid oxidation. The elements in the iron-constantan couple (iron, copper and nickel) have oxides that are easily reduced and thus there is less danger of parasitic surface effects. This couple maintains its accuracy until the metal is oxidized completely at one point of cross section whereas the first two couples may have an early failure. Protection tubes aid materially in decreasing the harmful effects of the surrounding medium on thermocouples. As yet there is no protection tube that is satisfactory for all applications. Therefore, the required characteristics of tube material are a function of the job.

Metallic protection tubes are usually suitable for base-metal couples at the lower temperatures, while a mullite or similar ceramic tube is used with Chromel-Alumel couples in the range of 2000 to 2400°F. Because of its tendency for rapid contamination, the platinum couple must be sealed gas tight in a dense ceramic tube impervious to the furnace atmosphere. Glazed porcelain or fused quartz are best suited for this purpose but are quite fragile. Therefore, they are used as inner tubes and protected from mechanical and thermal shocks with an outer tube of metal or ceramic material, depending on the temperature. If hydrogen atmosphere comes in contact with porcelain or quartz at high temperatures, the silica is reduced to silicon vapor which readily contaminates the platinum couple. Inconel tubes are widely used for protection around 2100°F. The thermal insulating effect of this double protection-tube assembly causes a definite heat-transfer lag between the actual furnace temperature and the platinum couple. The greater this lag the more sluggish and irregular is the furnace control. Assigning a temperature-lag index of 100 to a bare 14-gage couple, the over-all conductivities of the various types of protection tubes are shown in Table 2.

When the protection tube and couple are placed next to the material to be heated, the temperature variation may be considerable even though an automatic control mechanism is used. This variation is dependent, to a considerable degree, on the temperature lag between the heat input, elements or burners, and the thermocouple hot junction. Where minimum temperature variation is required, the couple should be located relatively close to the heat input, thus minimizing the temperature lag between the heat input and the controlling mechanism. With the couple adjacent to the work, good temperature control can be secured only if the heat-transfer lag between the input and couple is small. With some of the new controllers, however, in spite of a certain amount of lag, a practically straight-line control

TABLE 2

Types of Protection Tubes and Their Relative Transfer Lag

Material and Parameters	Relative Value	Material and Parameters	Relative Value	Material and Parameters	Relative Value
Bare couple, 14 gage	100	Inconel	29	Sillimanite, double tube	9
PYOD type	98	Seamless steel	26	Silicon carbide	18
Copper, thin wall	92	Chrome iron	26	Carborundum 1 in. internal diam.	22
Copper, heavy wall	86	Cast iron	20	Carborundum 2 in. internal diam.	10
Bronze	68	Nichrome	24	Mullite	12
Wrought iron ½ in. internal diam.	47	Fused quartz	30	Porcelain	14
Wrought iron 1 in. internal diam.	38	Graphite	36	Fire clay	7
Monel	36	Carbon	30	Porcelain and fire clay	3
Nickel	34	Vycor 96% silica	34	Sillimanite and carborundum	6
Stainless steel	31	Sillimanite	20		

can be secured. Where the lag may be quite large, the control couple is placed close to the source of heat, while the actual temperature of the charge is recorded by another couple adjacent to the charge. Therefore, to maintain the charge at a given temperature, the control mechanism will have to be set at a slightly higher temperature to compensate for the various heat losses between the heat source and the material to be heated.

For controlling the temperature of acid cleaning baths and similar applications, protection wells of compressed carbon are used successfully. They are quite impervious to and unaffected by acids and also withstand considerable handling. No protection tubes are required when iron-constantan couples are used to measure actual steel temperatures during bright box annealing. The annealing temperatures are approximately 1250°F. and the atmospheres are reducing in nature. Thus the sensitivity of recording actual steel temperatures is good without any detrimental effect on the life of the thermocouple. However, replacements must be made when the insulating beads become covered with carbon as the result of atmosphere and oil decomposition. This layer of carbon, if permitted to accumulate, will short-circuit the thermocouple millivolt output. The insulating beads may be cleaned by heating them in a furnace at 1600°F., thus burning off the carbon.

High-Temperature Thermocouples

For many years, there has been a concerted effort to improve the scientific control of steel making. One step in this direction is the reading of steel temperatures with ± 5° accuracy at 3000°F. Special thermocouples that have been used for experimental work include tungsten-molybdenum, tungsten-graphite and graphite-silicon carbide. These high-temperature thermocouples must have good protection from air and slag to prevent rapid deterioration. In the case of the platinum-platinum rhodium couple, the inner insulating tube should be in one piece so as to reduce contamination of the wire at the joining surfaces of the insulation beads. Platinum-platinum rhodium couples are successfully used in actual practice.

For measuring liquid steel temperatures, the thermocouple is mounted at the end of a long handle. This unit is immersed into the liquid steel by the observer. The lead wires extend through the handle to the couple. Since these couples are short and are placed inside the furnace, the cold junction must be cooled to obtain millivolt readings. This is accomplished by running water through the handle to a water-jacketed cold junction or, in some cases, by air-blast cooling. Prior to immersion, the slag is scrapped away from the small area in which the temperature is to be determined.

The tungsten and molybdenum wires are protected with

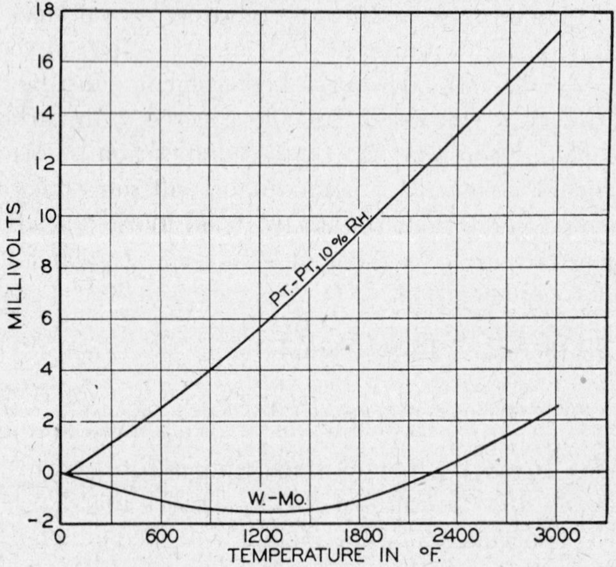

FIGURE 5. *Electromotive Force—Temperature Curve for the Tungsten-Molybdenum Thermocouple, also, Standard Platinum-Couple Curve*

an inner tube of sillimanite and an outer tube of sintered corundum. This couple, however, has a low millivolt output as illustrated in Figure 5. One immersion, to a depth of 16 inches, requires 2½ to 3 minutes and the life of the couple is approximately 6 heats. Each couple has to be individually calibrated and it is quite uncertain whether or not a given immersion would result in a satisfactory reading.

The tungsten-graphite thermocouple has the advantage of generating considerable electromotive force at steel-making temperatures, as shown in Figure 6. This thermocouple is composed of a tungsten wire of ⅛ inch diameter, suspended axially within a graphite tube and having a graphite plug in the bottom to hold the wire. An outer refractory tube protects the entire unit. The couple is quite fragile and must be also protected from reaction with the steel. A relatively

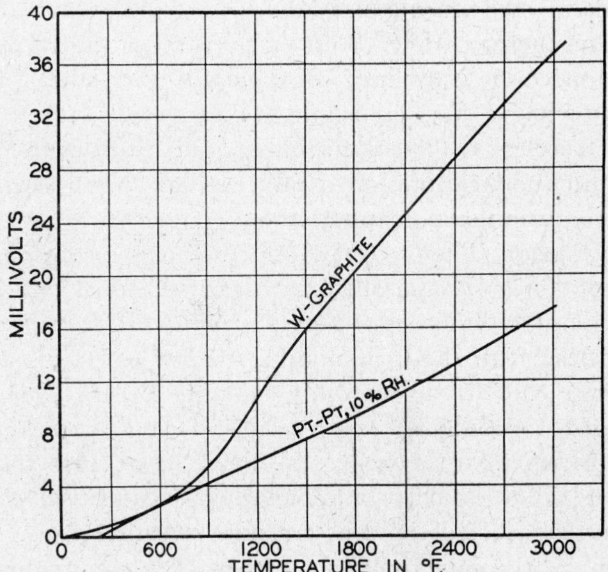

FIGURE 6. *Electromotive Force-Temperature Curve for the Tungsten-Graphite Thermocouple, also, Standard Platinum-Couple Curve*

large variation in cold-junction temperature can be tolerated so that the inconvenience and danger associated with water cooling can be eliminated. The calibration of these instruments is uncertain because of variations within the tungsten and moisture pickup in the graphite. Compensating leads of high- and low-carbon steel wire have been developed which permit the use of an air-cooled head on a portable unit.

The carbon-silicon carbide thermocouple, developed by Fitterer, has been quite widely used. The big advantage of this couple is its high millivolt value (approximately 0.5 volt) at 3000°F. This unit is constructed in a manner similar to the tungsten-graphite couple, with the silicon-carbide rod, $\frac{5}{8}$ inches in diameter, placed within the graphite tube. About the same advantages and disadvantages hold for this couple as for the tungsten-graphite unit.

Now we come to the old stand-by, the platinum-platinum rhodium thermocouple. Both the 10 percent and 13 percent rhodium couples are used as standards and other couples are rated for accuracy according to their deviation from these noble couples. If these platinum couples are kept free from contamination, calibration is unnecessary. When assembled in the conventional manner, these couples also have their disadvantages. They must be protected with a dense refractory tube of porcelain which in turn is sheathed in a heavy graphite tube. In order to keep the molten steel from rapidly reacting with the carbon tube, a wash of clay is added. However, the whole assembly is quite brittle and easily damaged. When the couple is immersed in molten steel, the great thermal shock may be sufficient to crack the protection tubes which would completely ruin the couple. Graphitic carbon is used as an outer sheath because of its higher heat conductivity than that of most nonmetallic refractory materials. Nevertheless, the couple is sluggish in its temperature response. In order to eliminate these difficulties, a quick immersion technique was developed. Figure 7

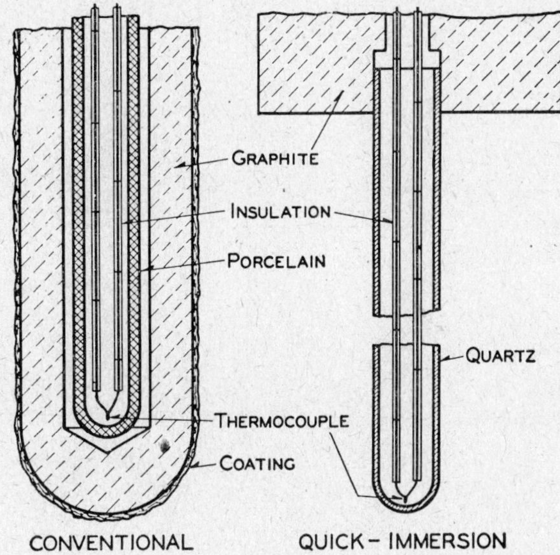

FIGURE 7. *Constructional Differences between the Conventional and Quick-Immersion Platinum Thermocouples for Molten-Steel Temperature Measurement*

illustrates the main constructional differences between the conventional and the quick-immersion couple. A thin quartz tube, about 3 inches long, is the only protection used at the hot junction of the couple. The low expansion coefficient of quartz permits its use at molten-steel temperatures without serious danger of breakage. Above this 3 inch zone, the couple is protected in the conventional manner. With this couple, the immersion time is 15 to 30 seconds and good checks are obtainable. The quartz tubes are replaced after each immersion.

Chapter 2

TEMPERATURE INDICATORS, RECORDERS, AND CONTROLLERS

Millivoltmeter-type Indicators

To measure, record, and control the temperature of a furnace, three interlocked circuits are required. The first consists of the thermocouple and lead wires, which produce a millivolt output varying in direct proportion to the furnace temperature. This millivolt output is applied to the indicating or recording mechanism. In this second circuit, the output of the thermocouple is translated into mechanical motion to indicate the temperature in degrees on a calibrated scale. In the third or controlling circuit, the mechanical action is transformed into electrical or pneumatic action which regulates the fuel valves or electrical contactors on the furnace.

Since the electromotive force of a thermocouple is calibrated against the temperature of its hot junction, an accurate means of converting this output into mechanical indication is all that is required to determine the temperature. Three types of indicating instruments are used for this purpose: millivoltmeters, direct-current potentiometers, and alternating-current bridge-type potentiometers.

A millivoltmeter, as shown in Figure 8, connected to the thermocouple, is the simplest and cheapest instrument for temperature measurement. However, it is sensitive to vi-

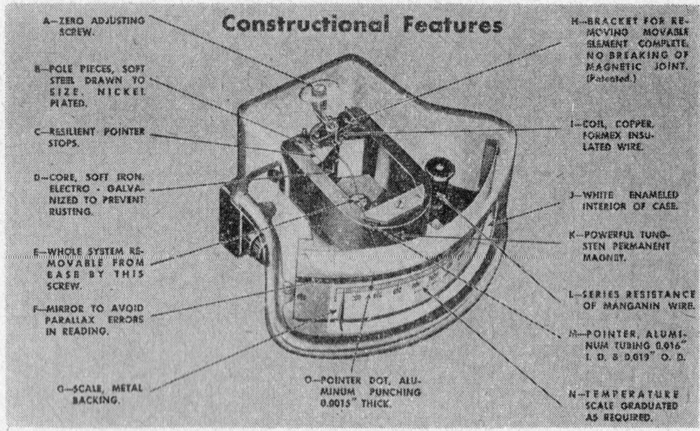

FIGURE 8.   *Diagramatic Sketch of a Temperature-Indicating Millivoltmeter*

bration, and the presence of dust particles around the movable coil will produce errors. If the length and diameter of the couple or lead wires are changed considerably, a correction must be made in the meter calibration. This adjustment is necessary to give the proper current flow through the galvanometer coil. The scale can be calibrated either in millivolts (requiring a millivolt-temperature conversion chart) or directly in degrees. This meter consists of four principal parts: a galvanometer and suspension, a permanent magnet, a temperature or millivolt scale, and a calibrated spool of resistance wire. The galvanometer is of the movable-coil type and is mounted between the poles of a permanent magnet. The deflection of the indicating needle, attached to the movable coil, is proportional to the generated millivoltage of the thermocouple. The current produced as a result of the thermocouple millivoltage passes through the movable coil and creates a magnetic field. This field tends to align itself parallel to the field of the permanent magnet. The opposing force here is a spring or a torque suspension which tends to balance the torque created by the reaction of the two magnetic fields.

If it is desired to change the type of thermocouple used or the temperature range of this instrument, the number of turns and size of wire used in the movable coil as well as the hair springs attached to the coil must be changed. The instrument is mechanically adjusted, on open circuit, by means of the adjusting screw which shifts the galvanometer needle. This needle is set so that it indicates the cold-junction or compensated temperature on the scale. However, electrical adjustment is also necessary to insure the correct relationship between the calibrated galvanometer and the current flow resulting from the millivoltage generated by the thermocouple. This is accomplished by adding or removing wire on the calibrating resistance spool which is in series with the thermocouple and galvanometer. If the resistance is increased by a change in thermocouple size, length of leads, or galvanometer coil windings, then the resistance of the calibrating spool must be decreased to maintain the correct over-all circuit resistance for proper calibration. In this thermocouple-lead wire-millivoltmeter circuit, Ohm's law is effective, that is, the voltage output of the thermocouple is equal to the product of the resistance and the current flowing in this circuit. Changing from one type of thermocouple to another (iron-constantan to Chromel-Alumel, for example) will require a change in both the galvanometer and calibrated resistance coils so that the temperature may be read directly on the scale. If no such changes are made, low readings will be obtained because of the lower millivoltage output of the Chromel-Alumel couple. If the scale is given in millivolts instead of degrees of temperature, no coil changes will be necessary.

This instrument should be mounted where vibration is kept at a minimum since the jewels, pivots, and suspension are quite fragile and prone to damage. No external source of current other than that furnished by the thermocouple is required to operate the millivoltmeter. Therefore, it is simple to install and quite compact. One use for this instru-

    Modern Pyrometry

ment is on ships to check diesel-engine temperatures; such a
unit is illustrated in Figure 9.

FIGURE 9.   *Bristol Millivoltmeter-Type Instrument for
Checking Diesel-Engine Temperatures*

The selector switch below the indicator enables several
couples to be checked individually on the same instrument.
With an open circuit, the indicating needle can be set at zero
by turning the outside adjusting screw. Another illustration
of the use of this type of instrument is the self-contained
portable unit made for instantly measuring surface temperatures of heated zinc, aluminum, or brass billets and steel
rails. With this instrument, various special couples may be
used and they may be replaced. The millivoltmeter should
be sturdily built to withstand handling, since it is an inter-
gral part of the thermocouple fixture. For a quick check of
the surface temperature of rails or billets, a thermocouple

having two sharp replaceable tips may be used. With this type of thermocouple, the wires at the hot junction are not welded together, but are separated. The hot junction in this case, is formed when the pronged wire tips contact the metallic heated body, the body thus becoming part of the hot junction. The temperature obtained is the average temperature between the tips. These tips are held in contact with the rest of the thermocouple by means of adapters and set-screws. Sufficient pressure must be applied, when checking temperatures, to have both points in good contact with the metal. For measuring surface temperatures of very small areas, an asbestos block with a small stainless steel button attached to one edge may be used. The two thermocouple wires are connected to the steel button, which heats very rapidly since the amount of heat conducted away by the asbestos block is quite small. For measuring roll-surface temperatures quickly, a special bow spring, holding a thin replaceable bimetallic ribbon couple, is pushed against the curved surface. Considerable care must be taken when

FIGURE 10. *Portable Hand Pyrometer with an Indicating Millivoltmeter and Replaceable Thermocouple Tip (Courtesy, Illinois Testing Laboratories)*

studying surface temperatures since poor contact between the surface to be checked and the element of the thermocouple may introduce serious errors. As shown in Figure 10, the millivoltmeter may be incorporated in the handle of the thermocouple unit.

With the millivolt instrument, fluctuations in the cold-junction temperature of the thermocouple must be compensated for. Since the voltage produced by the thermocouple is proportional to the difference in temperature between the hot and cold junctions, an increase in cold-junction temperature will produce a millivoltage drop of the couple, which is a false indication of a low hot-junction temperature. The compensation can be carried out by mechanically adjusting the zero position of the needle to the cold-junction temperature when the instrument is on an open circuit. The cold-junction temperature is obtained with a thermometer. On most of the meters, the cold-junction temperature compensation is accomplished automatically as illustrated in Figure 11, in which a bimetallic spring expands and contracts with

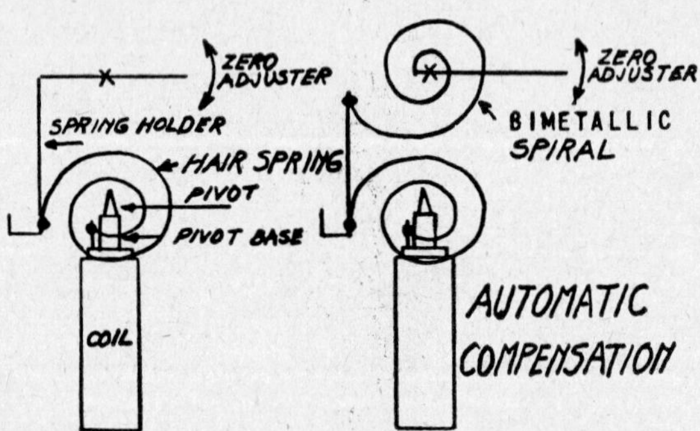

FIGURE 11. *Comparison of the Regular with the Automatic Temperature-Compensated Galvanometer Suspension (Courtesy, Brown Instrument Co.)*

the changes in cold-junction temperature. This movement automatically sets the indicating needle at the proper position. At the zero end of the scale, an indicator is mounted on another bimetallic spring with which the galvanometer is matched when the instrument is on open circuit, that is, it is disconnected from the thermocouple circuit. Once the needle is set in this manner, slight fluctuations in cold junction temperature will not affect the indicated temperature of the hot junction.

Potentiometer-type Indicators

The millivolt output of a thermocouple may also be measured by the null-balance method. The millivoltmeter must be able to measure the millivoltage of a continually flowing current accurately with no loss in the meter itself. This requires a high-quality millivoltmeter, having high internal resistance and sensitivity. Using the balancing technique, it is possible to indicate the temperature when the thermocouple millivolt output is exactly balanced by a known calibrated counter-direct-current output. With the balance type of instrument, variations in the external resistance of the circuit produce no harmful effects on the accuracy of the reading. The independence of measuring accuracy from the external line resistance is due to the fact that no current is flowing at the moment of balance. Prior to balance, a maximum current is flowing through the thermocouple circuit. By increasing an opposing direct current, balance of both is finally secured and the resultant current flow is zero. Thus, the resistance of the thermocouple circuit becomes infinity at balance. Any changes in the external resistance of the thermocouple circuit do not affect the actual point of balance. However, a high external resistance will make the balancing action sluggish.

The galvanometer, instead of actually indicating the temperature, acts as a null-point detector, so that small variations in the electromotive-force output of the couple are more easily detected. Figure 12 represents a simple potentiometer circuit, employing the balance method of millivolt measurement. In this circuit, the output of the thermocouple M is balanced against the battery B, using the slide-wire contact pointer F. The galvanometer G is used only as a means of indicating when balance is obtained between the thermocouple and the calibrated battery in the system. This mea-

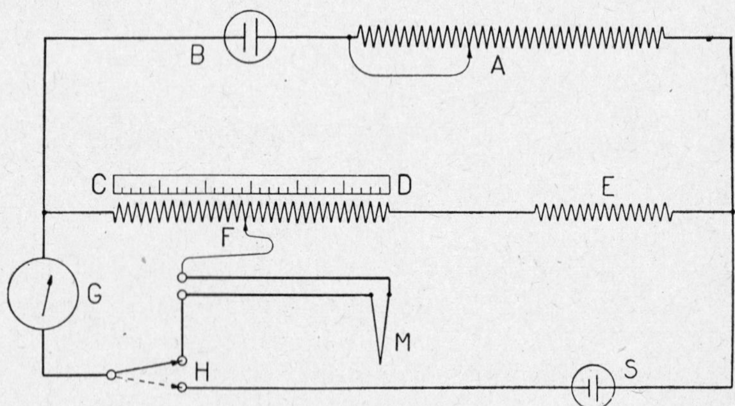

FIGURE 12.   *Simple Potentiometer Circuit—Null or Balance Method*

surement requires the accurate calibration or standardization of two important parts of the system, the slide-wire scale CD and the battery B. The slide wire can be calibrated either in millivoltage or in temperature degrees. When calibrated in degrees, only one type of thermocouple can be used per calibration since the temperature-millivoltage relationships of various couples are different. Any couple may be used with the millivolt calibration, but individual millivolt-temperature conversion tables are necessary. The temperature calibration is used on equipment that is permanently installed for furnace-temperature control and recording. Checking and testing require portable potentiometers that are calibrated with a millivolt scale. For accurate measurement, the dry cell B must be able to supply a known potential to the balancing circuit. This constant dry-cell output is secured through calibration with the standard cell S. By moving the switch H to the dotted position, the standard-cell millivoltage is opposed to that of the battery, the thermocouple being disconnected. The variable resistance A is then adjusted until the galvanometer indicates balance, so that the millivolt outputs of the battery and standard cell are identical. The standard cell is the Weston or cadmium type and yields a standardized electromotive force of 1.0183 volts at 20°C., with a low temperature coefficient. This standard cell is in the circuit for just the short period of time required to standardize the battery output, therefore, it has quite a long life.

FIGURE 13. *Operation of the Balancing Mechanism of the Leeds and Northrup Micromax Unit (Courtesy, Leeds and Northrup Co.)*

If the millivolt output of the thermocouple changes due to a drop in temperature, the galvanometer is deflected, indicating that the slide wire must be moved to rebalance the circuit. The dry-cell output is greater than that of the thermocouple. The adjustment of the slide wire is effected and electrical balance is maintained automatically in the motor-driven instruments, one type of which is illustrated in Figure 13. This figure is a view of the balancing mechanism of the Leeds and Northrup Micromax unit. Section 1 shows the electrical system in balance. The galvanometer needle extending forward between the clamp bars is in the center position indicating that the current from the thermocouple equals that from the dry cell. Section 2 shows the two vertical setting levers closed on the galvanometer needle in the zero position when no change is transmitted to the slide-wire balancing mechanism. Section 3 indicates that the millivoltage has changed, unbalancing the system and displacing the galvanometer needle to the right. The setting levers are open giving maximum galvanometer freedom. Section 4 illustrates more fully the component parts used in transmitting the galvanometer displacement to the balancing slide wire. The initial operation is to temporarily clamp the galvanometer so that its relative position may be determined by the sensing fingers. This is accomplished by the upward movement of the horizontal bar just beneath the galvanometer needle. The sensing fingers are crossed similarly to a pair of scissors, and are cam operated with a small return spring. The cam opens the fingers at the top, then the return spring pulls them together to contact the positioned galvanometer needle. When the needle is to the right of the zero position (as illustrated) the left finger closes through a greater arc than the right finger. Therefore, the lower portion of the left finger must swing through an equivalent arc to the left. In doing so, it hits the eccentric pin on the clutch bar, causing it to rotate counterclockwise. Proper magnification of the galvanometer needle movement is secured by making the lower portion of this sensing finger path longer than the upper portion. The front assembly containing the sensing-finger and clutch-bar mechanism has bearing supports at the top to permit the bottom to swing forward or backward. This swing action, energized by cams and return springs, occurs periodically about every 2 seconds. The sensing-finger action occurs when the bottom of the front assembly is pushed forward. Upon its return, through spring action, the clutch bar makes pressure contact with the clutch plate. With this bar tilted and firmly clamped the two large cams on the horizontal shaft rotate counterclockwise to push the displaced clutch bar clockwise back to the horizontal position. This produces a clockwise rotation of the clutch disk and, being on the same shaft, the slide wire is rotated an equivalent amount. As the slide wire is rotated and balance is approached, the galvanometer moves to the zero position. Balancing continues by increments until the action is complete. The galvanometer coil is fully suspended, using a fine noncorroding wire. For indicating or recording, an indicating pointer or pen is connected to the slide-wire balancing mechanism.

Since the amount of electromotive force generated by the thermocouple is proportional to the difference between the temperatures of the hot and cold junctions, it is necessary that the reference (cold) junction temperature is held constant. Otherwise, proper compensation must be made for any temperature variations at the cold junction. This is accomplished by inserting a small coil of resistance wire, of proper temperature coefficient, into the balancing circuit (resistance E in Figure 12). If the hot junction is maintained at a constant temperature, the position of the temperature indicator should remain constant even though the reference junction temperature may fluctuate due to room-temperature changes. When the temperature of the reference junction increases slightly, the resulting effective thermocouple millivoltage is actually lowered since the temperature difference between the hot and cold junctions has been reduced. However, the simultaneous increase in resistance of this compensating coil, due to a rise of the surrounding room temperature, balances the reduction of effective thermocouple millivoltage. Thus the increment of resistance increase in this particular branch of the potentiometer circuit is equivalent to the loss in millivolt output of the thermocouple. Therefore, the indicated temperature will remain the same even though the millivolt output of the couple has actually decreased.

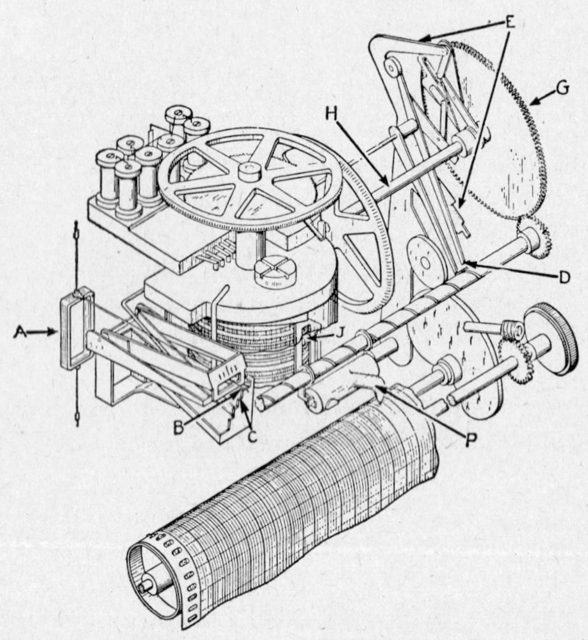

FIGURE 14. *Detecting and Rebalancing Mechanism of a Conventional Galvanometer-Type Potentiometer (Courtesy, Brown Instrument Co.)*

The direct current supplied to balance the variable thermocouple output is obtained from a dry-cell battery. To keep the dry-cell battery output standard, it is periodically balanced against the output of a standard cell through temporary connection and current adjustment with a small rheostat. In manually balanced instruments, a button is pressed to properly connect the standard cell. Then the rheostat in the battery circuit is adjusted until no deflection is indicated by the galvanometer. Automatic balancing mechanisms are now being used which connect the standard cell at given intervals (about every 40 minutes) and adjustment is made automatically as the instrument runs in normal operation. As contact is made with the standard cell, the instrument balancing action engages the standardizing rheostat and rotates it till the galvanometer goes to the zero position, indicating the battery and standard cell to be at the same potential.

The Brown balancing mechanism, utilizing a step selector, is illustrated in Figure 14. In this instrument, the degree of off-center movement of the galvanometer B is determined by means of a calibrated step selector C. This selector is moved upward once every 3.6 seconds to carefully contact the galvanometer needle. A light clamping action on the galvanometer precedes this positioning of the step selector. When the needle is over to the right, the selector only rises a short distance before one of the steps clamps the galvanometer in position. When the galvanometer is displaced to the left, there is more vertical travel of the step selector before contact is made. The steps on the selector are designed to yield maximum sensitivity when the galvanometer needle is close to the zero position. This permits any slight displacement of the needle to be measured by the selector. This step selector, by translating the degree of galvanometer unbalance into varying amounts of vertical motion, actuates a secondary positioning pointer D. A kidney-shaped rotating cam and a spring return action cause another step selector E to check the position of pointer D. The forward motion of E stops when it contacts a pin located at the side of pointer D. The position of step selector E is now a magnified indication of the degree of balance of the galvanometer. The upper portion of E also moves forward an amount indicated by the lower portion. From this selector position, the gear G is activated to move the slide wire contacts to a balancing position. The direction and the degree of rotation of gear G are determined by the selector E for proper balancing of the circuit. By being submerged in oil, the spiral slide wire is kept free from dust and dirt while contact is made by means of slide J which moves around the periphery of the slide-wire coil. The movement of gear G also positions the recording pen P.

By offsetting the mechanical from the electrical zero point, a thermocouple burn-out safety feature is obtained. A broken or burned couple allows the galvanometer to assume its mechanical zero position which is displaced from the point of electrical balance. This moves the pen or indicator upscale above the set temperature on the control and shuts off the furnace.

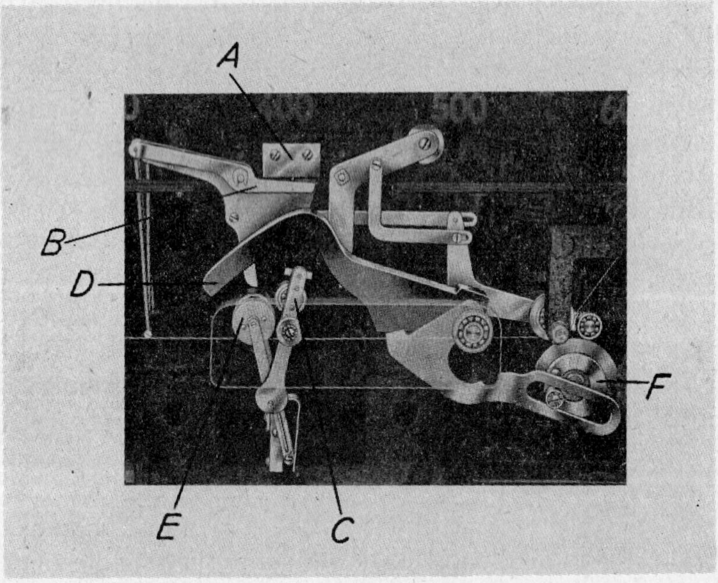

FIGURE 15. *Sensing Mechanism of the Foxboro Potenti-ometer (Courtesy, Foxboro Co.)*

A third type of balancing mechanism is illustrated in Figure 15 which indicates the operation of the Foxboro unit.

This unit transmits the position of the galvanometer to the controlling slide wire in still a different manner. The galvanometer needle is clamped in position against a plate A by the oscillating sensing-cam arrangement B. The upper edge of cam B has a calculated slope so that as the galvanometer needle moves from right to left, the angular motion of the cam diminishes. The lower portion of cam B extends down to move the friction roller C off center. When the galvanometer is to the left, the friction roller C moves only a short distance, and when the recentering device D moves downward, roller C rides up to the top of the inverted V. The extreme top of this inverted V-arm is the point of balance of the galvanometer needle. More displacement of friction roller C (always to the right) occurs when the galvanometer needle is at the right and recentering moves the roller back in the opposite direction to the neutral or balanced point. At balance, the galvanometer will cause the friction roller to move to the right until it coincides with the top of the inverted-V recentering arm. This condition will produce no movement of the friction roller. The centering of roller C causes roller E to drive the slide-wire contact and pen carriage in the proper direction for balancing. With the galvanometer in a position to the left of center, friction roller C is moved only a small amount and recentering device D will produce still further movement to the right as the roller moves up the arm to the neutral position. Therefore, the pen carriage and slide-wire contact move up- or downscale depending upon which side of the inverted V-arm is used for centering. The cam at F drives this balancing mechanism.

All these balancing mechanisms are designed to measure the displacement of the galvanometer needle, utilizing mechanical levers. In the Tag Celectray, however, the galvanometer movement is translated to the balancing slide wire by means of a light beam, a mirror-type galvanometer and a photoelectric cell (cesium-coated, vacuum-type). In this photoelectric circuit, there are also two relays which actuate a reversing balancing motor. When both relays are closed, the balancing motor moves the slide-wire contact downscale. To bring the motor to rest position, the first relay must be closed while the second remains open. The upscale movement is accomplished when both relays are open.

The potentiometer circuit is the same as normally used, but a mirror is mounted on the galvanometer coil instead of a needle. This mirror turns as the hot-junction temperature of the thermocouple is increased. Any galvanometer movement is detected by means of a small, focused, constant beam of light, hitting the mirror. This beam is then reflected back to the center line of the horizontal photoelectric cell. Therefore, any movement of the galvanometer yields a magnified horizontal movement of the reflected beam of light. At one end of this cell is an adjustable metal shield or "controlling edge" which sharply blocks out a section from the reflected light beam.

A rise in the thermocouple temperature will cause a deflection of the galvanometer so that the light reflected from the mirror will be completely blocked off from the photoelectric cell by the metal "controlling edge," thus permitting both relays to open. This condition energizes the balancing motor to drive the slide-wire contact and indicator upscale. As balance of the potentiometer circuit is secured, the light beam from the galvanometer shifts over until it is split by the "controlling edge"; that is, half of the light falls on the cell while the other portion is blocked out. This condition kicks in relay No. 1 (relay No. 2 being still deenergized) to stop the balancing action of the motor at the proper position of the slide-wire contact.

To yield this split-beam control, the action rates of the two relays are the opposite of each other; that is No. 1 opens slowly and closes fast while No. 2 opens fast and closes slowly. If the galvanometer moves as a result of vibration, a "back voltage" is generated. This voltage is the result of the closed galvanometer coil cutting the magnetic field of the surrounding permanent magnet. To improve the galvanometer stability and sensitivity, this back electromotive force generated by any movement of the galvanometer is eliminated by a bucking electromotive force supplied to the galvanometer as the balancing motor starts rotation.

A balancing system, utilizing the galvanometer in the capacity of a reversing-contact switch, has been developed by Bristol and its schematic diagram is shown in Figure 16. This particular instrument is the circular-chart type. It consists of five integral units: a potentiometer, a battery, a power pack, a galvanometer, and a relay.

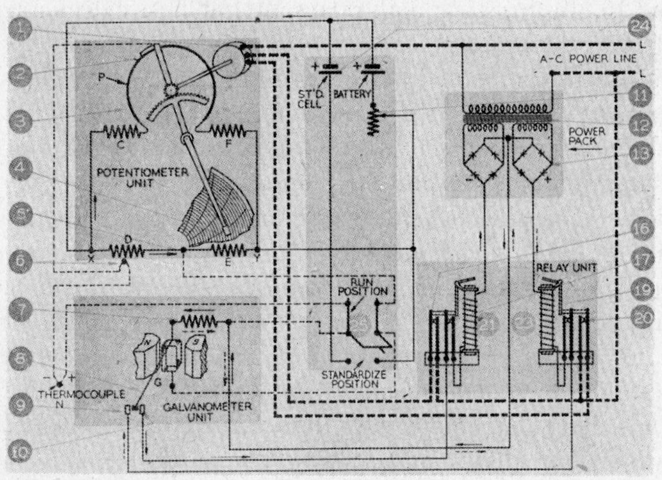

FIGURE 16.   *Principle of Operation of the Bristol Pyromaster Potentiometer Unit (Courtesy, Bristol Co.)*

This potentiometer circuit has a circular slide wire on which the sliding contact 2 is driven by means of a reversible motor 1. This motor is also geared to the recording pen. The automatic cold-end compensation is accomplished by the temperature-responsive wire at 6.

Standardization of the battery in the potentiometer circuit is accomplished similarly to the other instruments through use of the standard cell.

By means of "bridge" connected rectifiers, the power pack acts to convert alternating current into direct current of the proper polarity for operating the relays. These relays, in turn, make the circuit required to drive the balancing motor.

The galvanometer is connected to the thermocouple circuit (light dashed lines) in the usual manner. However, the galvanometer needle itself is ingeniously connected into the relay circuit to act as a control switch for the balancing motor. The tip of this needle has two contact points mounted on it. When the system is in balance, the galvanometer needle, in the null position, is midway between the two contact points 9 and 10. In this state, both relays, 21 and 23, are deenergized and the motor-circuit contacts 16 and 20 are both closed. This means that both the forward and reverse windings of the motor are energized, but being equal opposing forces, no movement occurs. As the temperature rises, the additional millivoltage from the thermocouple causes the galvanometer needle to be displaced and touch one of the contacts, 10. The initial touch may be very light, but when it occurs, a small current flows through resistance 7, galvanometer needle 8, contact 10 and relay 21. This current yields a potential drop across resistance 7, which is also connected in the galvanometer circuit (light dashed lines). This will permit more current to flow from the battery through the galvanometer circuit and, having the proper polarity, will force the galvanometer needle against the contact more firmly. Increasing the force will decrease the contact resistance at this point, thus allowing more current to flow. This will yield an increasing potential drop across the resistance 7. As the contact pressure builds up, the current is increased to the point where it is sufficient to operate the relay 21. Pressure build-up at the contact and resulting current flow takes place almost instantaneously. When relay 21 is energized, switches 16 and 17 are opened. Switch 16 opens the reverse-drive circuit of the balancing motor, thus allowing the forward motor drive to move the pen upscale. The opening of switch 17 at the same time, however, deenergizes the relay circuit to the galvanometer, thus stopping the balancing motor. The relay becomes reenergized immediately as switch 17 is closed. This rapid alternate action of the relay continues until the driving motor moves the slide-wire contactor to a balance position with the thermocouple voltage output and the galvanometer goes into the neutral position. A decrease in temperature of the thermocouple hot junction causes a similar action to take place with contact 9 and relay 23. The contacts 9 and 10 are set quite close to the galvanometer needle so that a very slight galvanometer motion is sufficient to make contact.

Electronic Balancing Mechanisms

Modifications in the potentiometer circuit secure continuous balance through the use of electron tubes. The modified instruments give very rapid temperature recordings since pen speeds of full-scale travel in $1\frac{1}{2}$ seconds or less are possible. These instruments have also been successfully used for the rapid recording of applied stresses in airplane structural members as detected by the SR-4 type strain gage. The electronic instruments have fewer moving parts and can be operated under conditions of vibration and shock which would ruin the standard galvanometer suspension, but they are somewhat more expensive and require the services of a mechanic well versed in electronics.

The Leeds and Northrup Speedomax Type A instrument is one of the extremely high-speed temperature recorders and has been applied to quenching-speed recording and microphotometer measurements in spectrography. The electronic balancing of a circuit, in general, is accomplished by taking the output millivoltage of the thermocouple (direct current), converting it into a fluctuating voltage and amplifying it in an electrical circuit to actuate the slide-wire driving motor either forward or reverse. In this particular unit, the galvanometer has been eliminated from the balancing circuit, but is required in the standard cell circuit. A microphone drive, constantly energized by 60-cycle 110-volt alternating current, converts this driving energy into mechanical vibrations of similar frequency. The motions of the oscillating plate in this unit are transmitted by mechanical linkage to a carbon microphone similar to that of a telephone transmitter. The thermocouple, balancing battery and slide wire as well as the primary of a transformer are series connected to the microphone chopper. The changing pressure on the carbon granules, due to the mechanical vibrations, produces variations in their electrical resistance so that any direct current which passes through them will be given a sine-wave pulsation. The direction of the flow of this current depends upon whether the thermocouple millivoltage is higher or lower than that of the potentiometer circuit (similar to the mechanically balanced potentiometer). The direct current has been given a pulsating characteristic so that it may be handled by a transformer and then fed into an electronic circuit. This variation in intensity is sufficient to produce changing lines of magnetic flux in the primary winding of the transformer and thus to induce an alternating current in the secondary. The secondary output of the transformer is amplified and passed through a second transformer having two secondary windings to which are connected the grids of two thyratron tubes of opposite phase. The grid of a thyratron tube is located between the cathode and anode and acts as an electronic switch on the current flow in the tube. An electrical potential is maintained between the cathode and anode, but no current flows unless the grid has the proper alternating-current phase. The direction of current flow, either from the thermocouple or from the balancing battery, determines this phase. If the alternating-current wave form of the grid voltage is opposite of that of the potential applied between the anode and cathode, no current flows in the circuit. But when the wave forms of both voltages are the same (in phase) current flows through the tube.

The outputs of these tubes are connected respectively to the forward and reverse windings of a three-lead split series reversing type of motor which drives the balancing slide wire of the potentiometer.

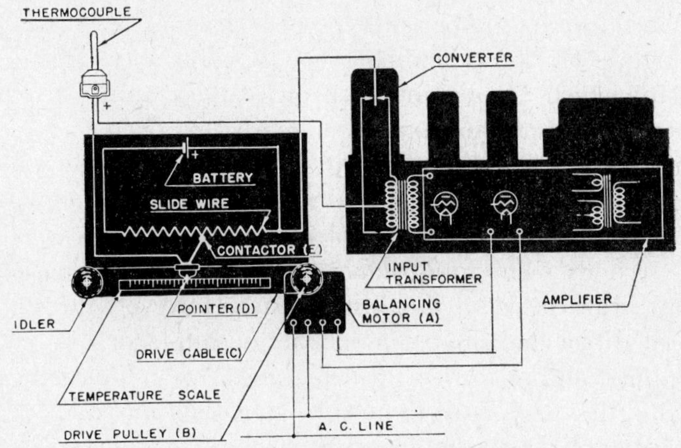

FIGURE 17. *Diagram of the Brown Electronik Potentiometer (Courtesy, Brown Instrument Co.)*

When the millivoltage output of the thermocouple is greater than that of the dry cell in the potentiometer circuit a small current flows through the microphone chopper to the input transformer. This alternating-current component is amplified to operate a thyratron tube of the proper phase which drives the slide-wire balancing motor forward to a new position indicating the higher temperature. If the thermocouple temperature falls, the small current flows from the potentiometer battery. This opposite current flow, a phase shift of 180 degrees, will then actuate the other thyratron, thus driving the slide wire and indicator downscale to the new temperature. As the null or balance point is reached, no current flows between the thermocouple and potentiometer. Therefore, there is no action in the input transformer since the rate of change of flux is zero.

The tendency of this high-speed balancing motor to overshoot after a very rapid temperature change is eliminated by means of a magneto tachometer attached to the drive shaft. The output of this source, increasing directly with the motor speed, opposes that of the thermocouple circuit and tends to bring the other thyratron into action to slow down the series motor as the point of balance is reached.

Figure 17, shows another type of continuous balancing circuit which is used in the Brown "Electronik" potentiometer. In this circuit, the galvanometer has been completely eliminated, thus decreasing the number of moving parts and avoiding any effects of vibration. The maximum full-scale pen speed for this type of instrument is approximately $4\frac{1}{2}$ seconds. The null-balance method of voltage measurement is also used in this instrument. However, the unbalanced direct current from the thermocouple or dry cell is amplified and measured, utilizing an electronic circuit. To do this, the direct current must be converted into alternating current. This is accomplished by means of a converter which is similar to that used in automobile radios.

The converter or chopper in this instrument is a flat-metal vibrating reed which is actuated by a 110-volt, 60-cycle driving coil. The reed, being alternately magnitized 120 times a second, moves back and forth between the poles of a permanent magnet. The unbalanced direct-current milli-voltage passes through the reed alternately to the two contact points and then to the primary side of the input transformer. A center tap on the primary of this transformer, connected to the potentiometer circuit, is necessary to utilize the full converter action. During one-half cycle of the converter operation, the unbalanced direct-current milli-voltage flows through one-half of the primary transformer winding, and then, for the balance of the converter cycle, this current is rerouted to the other portion of the primary winding to yield a complete reversal of magnetic flux. This action converts the unbalanced direct-current millivoltage into 60-cycle alternating-current millivoltage in the secondary of the transformer, thus permitting the use of an electron-tube ampifier and control circuit.

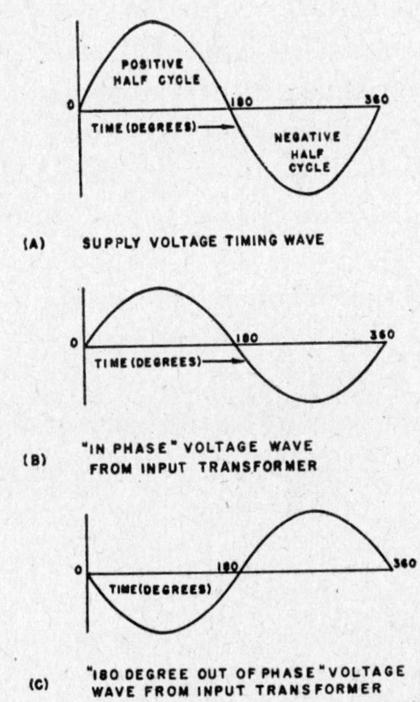

FIGURE 18.   *Voltage Wave Characteristics Utilized in the Brown Electronik Balancing Circuit (Courtesy, Brown Instrument Co.)*

The forward or reverse motion of the balancing motor is dependent on the phase relationship between the voltage that is being converted and that which is driving the converter. As the temperature rises in the furnace, the millivoltage output of the thermocouple increases to unbalance the potentiometer circuit. This unbalanced direct-current millivoltage is then converted into an alternating voltage of proportional magnitude. The converter, in which this change occurs, is driven by a 60-cycle supply voltage which has a wave form illustrated in curve A of Figure 18. This unbalanced thermocouple voltage, after passing through the converter, pulsates in the secondary of the input transformer as shown by curve B. The timing of the peaks in curve B

is identical to that in curve A. Consequently, the amplifier
drives the balancing motor in the proper direction. If the
furnace temperature falls, the polarity across the converter
and input transformer (Figure 17) reverses, since the potentiometer battery becomes the source of potential applied to
the reed converter. Then a voltage wave, similar to that in
curve C, is produced. In this case, the voltage lags behind
the supply-voltage timing wave by 180 degrees and is considered "out of phase." This condition, when properly
amplified, causes the balancing motor to operate in the reverse direction.

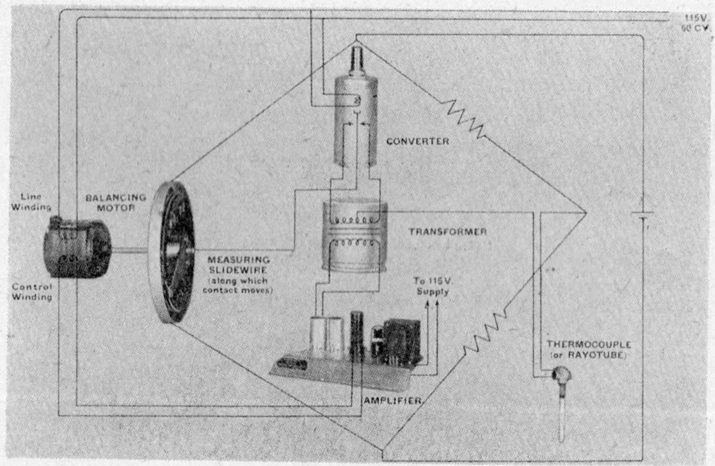

FIGURE 19. *The Essentials of a Type G Speedomax Potentiometer (Courtesy, Leeds and Northrup Co.)*

In this circuit, no galvanometer is used to determine the
degree of unbalance. The output of the thermocouple is
opposed to that of the balancing dry cell in normal operation. For standardizing, this dry-cell output is disconnected
from the thermocouple circuit and connected to the standard
cell. If the outputs of the cells are not matched, the resulting flow of current is amplified by the circuit to rotate the
balancing motor. When standardizing, the balancing motor

is engaged so as to move a rheostat in the proper direction till balance between the two is secured. The direction of current flow determines the direction of the balance-motor rotation. In this instrument, standardization takes place automatically every 30 minutes.

The Leeds and Northrup Speedomax Type G instrument operates in a manner similar to the Brown instrument in utilizing an electrically driven converter to make possible the electronic measurement of the unbalanced direct-current millivoltage. A schematic diagram of this circuit is shown in Figure 19. When used as a multiple-point recorder, as many as 16 points can be printed in as little time as 1 minute.

A new electronic recorder and indicator, that has been designed recently by Foxboro and called the "Dynalog," operates completely from vacuum tubes in amplifying, rectifying and filtering circuits, (Figure 20). In this instrument,

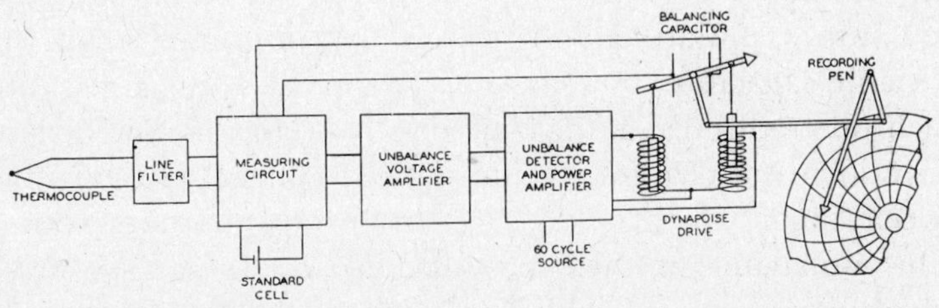

FIGURE 20.   *Block Diagram of the Dynalog Electromotive-Force Circuit (Courtesy, Foxboro Co.)*

the slide wire has been eliminated and in its place, there is a friction-free, rotating, variable air capacitor, similar to that used for tuning a radio. The continuous rotating motor is supplanted by a double solenoid-type drive that is directly coupled to the balancing capacitor. The galvanometer and the dry-cell battery have been discarded. These changes became possible by replacing the direct-current potentiometer circuit with a special alternating-current bridge-type circuit.

Two capacitors are used in this bridge, one being fixed and the other of the variable type. These replace the fixed resistor and the slide-wire resistance used in the direct-current potentiometer circuit. The extremely high resistance of the vacuum-tube measuring circuit enables the standard cell to be used directly since the current drain is about 0.1 microamperes. Thus, any additional periodic standardization of the dry-cell current in the circuit is unnecessary. Any change in the thermocouple direct-current millivoltage unbalances the alternating voltage in the vacuum-tube measuring circuit which, in turn, is amplified about 100,000 times. Such power is sufficient to actuate the proper driving solenoid for rebalancing the circuit through the variable capacitor.

This measuring system uses two capacitors for comparing the voltage from the thermocouple with that of the standard cell. By means of a vibrator, driven by 60-cycle current, the output of the thermocouple is impressed on the fixed capacitor. A similar vibrator converts the output of the standard cell into a pulsating millivoltage and impresses it on the variable capacitor. These two capacitors are discharged into each other through a high-speed switch, and if they do not have identical charges, an unbalanced voltage will exist between them which is detected by a vacuum-tube circuit. This unbalance is then amplified to the Dynapoise drive which rotates the variable capacitor till balance is secured between it and the fixed capacitor.

If the temperature at the hot junction of the thermocouple is less than its reference-junction temperature (for low-temperature measurements), the charge on the fixed capacitor will be reversed. Since it is impossible for the balancing condenser to take a zero or negative charge, a second condenser is charged by the same standard cell, but connected in a reverse direction. Therefore, balance is secured when the charge on the thermocouple capacitor is equivalent to the algebraic sum of the variable capacitor and its opposite-polarity neighbor. This added capacitor can be adjusted to set the reference point of the instrument.

Since these capacitors are operated from 60-cycle vibrators, the unbalanced voltage at the detector is of a pulsing nature. A timing contact is inserted into the amplifier circuit to cut out the capacitors during charging and switching. A filter is placed in line with the thermocouple so as to eliminate any possible pick up of 60-cycle stray current which would affect the operation of this 60-cycle instrument. This, together with the high internal resistance of the instrument, makes it possible to disregard lead-wire length and location, particularly with respect to other electrical apparatus.

FIGURE 21: *Solenoid Driving Mechanism for the Dynalog Instrument (Courtesy, Foxboro Co.)*

Figure 21 is a more detailed illustration of the driving mechanism. Connections are made to the pen arm for direct action on the circular chart. The moving portions of this type of drive are two iron rods which operate vertically within the two opposed solenoids. The dashpot aids in producing a smooth pen motion with no overshooting. When this system is in balance, the current in both solenoid windings is the same and there is no movement of the iron cores. But as soon as the thermocouple detects a rise in furnace temperature, one of these solenoids is supplied with more current than the other one. This stronger solenoid, overbalancing its partner, pulls its iron core downward and produces a rotation of the balancing capacitor rotor to a new balance position. The solenoids are designed to give the proper core movement for rebalancing regardless of the position of the core in the coil. This is accomplished by giving the cores and coils such dimensions that the cores are in the constant pull zone of the coils.

Automatic Recording

For a large number of applications, a continuous temperature record is required, especially for checking annealing cycles, heat treating and hot rolling. A complete record of an annealing cycle is extremely helpful in observing the actual temperature variation of the charge. Deviations in any number of cycles can be compared and reflected back to the quality of the products. If the temperature overshoots 25°F. during the spheroidizing anneal of high-carbon, cold-rolled strip, hard spots will occur. This treatment of high-carbon steel causes the iron carbide particles to agglomerate into balls or spheroids to yield a soft structure. Overshooting would cause localized transformation which would result in spots of harder pearlite on cooling. If the top temperature is too low in this operation, the carbide particles will not have sufficient mobility to coalesce rapidly. Therefore, for a given length of annealing cycle, the steel will not be fully softened. A quick glance at the recorded temperature cycle will show which of these conditions is present if the product is hard, and proper corrections can be made on subsequent charges. Also, by following the chart, the proper rate of heating can be given to hardening steels. For large steel sections, a slow heating rate is required to prevent the development of thermal stresses, which would cause distortion and cracking. With this steel section close to a recording thermocouple, its transformation from ferrite plus carbide to austenite (carbide in solution) can be followed on the chart and sufficient holding time given to insure complete uniformity prior to quenching. This uniformity is a requirement if best physical properties are to be obtained.

A needle-point scratch on a piece of smoked paper was one of the first means of securing a printed record. Photographic means have been used, but for modern industrial work, there are two general methods which are employed almost exclusively. They are the pen type, using fluid ink and making a continuous line, and the printing type, making a periodic dotted impression on the chart and using a multi-colored ribbon or an ink pad.

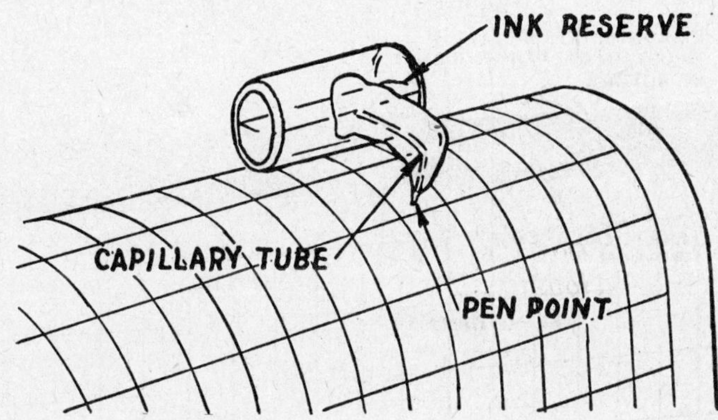

FIGURE 22. *Recording on a Roll Chart, Using a Fountain Pen—Continuous Line (Courtesy, Brown Instrument Co.)*

The fountain pen, as illustrated in Figure 22, carries a large supply of ink which flows out of the capillary tube down through the pen point. This unit is mounted in a holder which is driven up and down the scale, along with the temperature-indicating pointer. A good grade of chart paper as well as proper ink and pen are necessary to prevent the ink from spreading or blotting. If the ink flow is too rapid or the drying time too long, running and smearing will result. In some instances, rapid temperature fluctuations may cause blotting due to the closeness of the inked lines. This condition can be remedied by increasing the chart speed. In some cases, a more rapidly drying ink may be used. The pens should be checked daily to prevent clogging and running dry.

For application on circular charts, a bucket pen is attached to the bottom of the temperature-indicating arm. A small, rectangular metal bucket holds the ink, and a vertical capillary tube brings the ink up to a horizontal pen which contacts the paper chart.

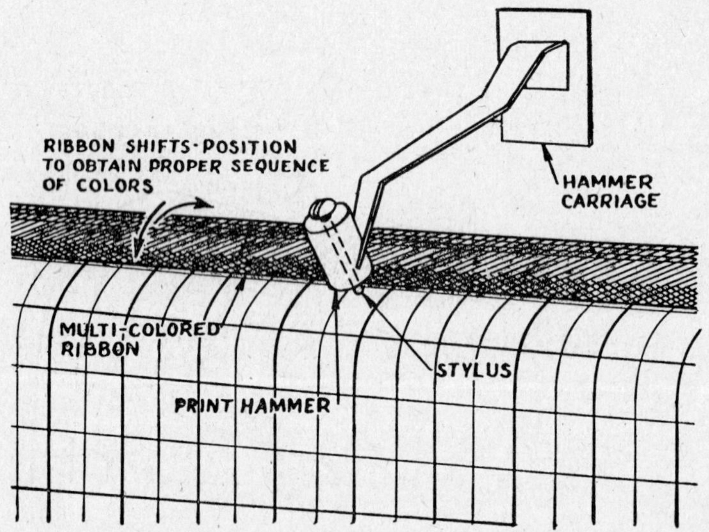

FIGURE 23. *Print-Hammer Recording—Evenly Spaced Impressions (Courtesy, Brown Instrument Co.)*

This continuous-line recording yields a complete picture of a given operation. However, for multiple recording on the same instrument, the color-printing technique is used. In Figure 23, a three-colored ribbon is used to give better separation of the recorded temperatures. As the recording instrument becomes balanced, the hammer carriage moves the hammer downward and, at the same time, the ribbon moves forward under the hammer to the proper color. This typewriter action produces a round colored impression. Three individual records can be secured by this means on a single chart. To record and separate as many as twelve

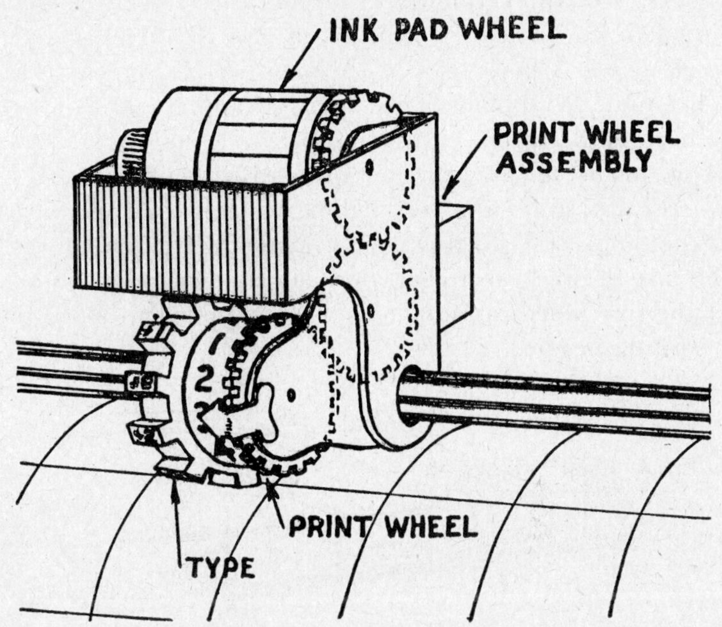

FIGURE 24. *Print-Wheel Mechanism for Multiple Recording (Courtesy, Brown Instrument Co.)*

individual processes on one recording instrument, a device similar to that illustrated in Figure 24 is used. The regular recording mechanism moves the print wheel up or down the scale on a slotted shaft. At the balance point, this shaft rotates to force the type wheel on the paper chart. The type wheel gives a dot or a plus-sign type of impression for accurate temperature indication. This mark is also identified with a printed number and color. After an impression is made, the print wheel rotates to force the type against the properly colored ink pad while moving to the next position.

This typewriter technique has been applied to millivolt-meters to secure a permanent record, (see Figure 25). A mechanically driven depressor bar periodically moves down to force the galvanometer needle against the ribbon and on the paper chart. With special high-speed recording milliam-meters used with certain types of photoelectric-cell tempera-ture-detecting systems, a continuous line record is obtained by utilizing a power-driven galvanometer as a pen. This horizontal galvanometer needle is a capillary tube which transports the ink from a large open well, located near the pivot, to the curved pen tip. This tip curves downward to a vertical position and very lightly contacts the paper chart at the top roll. With this arrangement, the friction between the pen and the paper chart does not influence the accuracy of the instrument as ample power is supplied to move the recording needle.

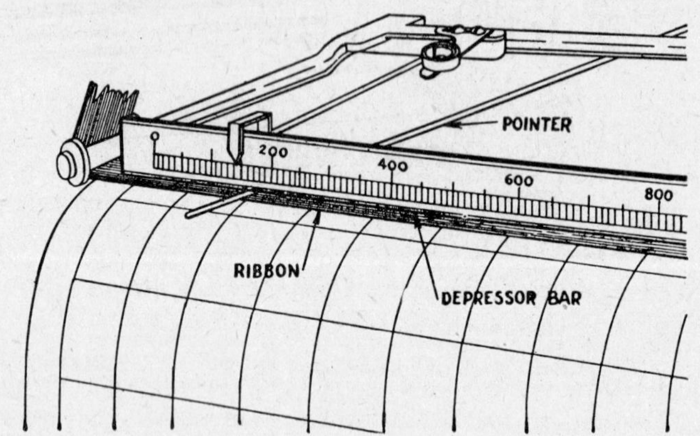

FIGURE 25.   *Depressor-Bar Printer for Millivoltmeter-Type Instruments (Courtesy, Brown Instrument Co.)*

In the rectangular chart instrument, a roll of paper is used. This paper is properly sized to take the ink and is ruled for temperature degrees across the width and for time along the length. This permits a study of the chart with respect to both time and temperature. The edges of the chart are perforated so that it is constantly and uniformly driven by the top sprocket roll. The pen or print hammer contacts the chart on this top roll. Friction pay-off and take-up spools keep the chart taut to insure accurate recording. With the circular-type instrument, the chart is a circular paper disk with a hole in the center. This chart is driven by the center shaft rotating at a fixed speed. It is held in position on the shaft by a pressure plate and nut. This type of chart is usually renewed every 24 or 48 hours, depending on the instrument speed. The temperature lines are circles increasing in radius from the center, while time is measured by a series of curved lines running radially from the center. This type of chart, although requiring daily replacement, permits the cycle of a given operation to be viewed at a glance.

Automatic Temperature Control

Depending on the type, size and characteristic temperature lag of a furnace, various methods of automatic control are used to secure a minimum temperature fluctuation about the control point. Any type of automatic temperature control requires an instrument that can be set at the desired temperature and will adjust the power input to bring the furnace up to this control temperature and maintain it within the desired limits of temperature variation.

A large heat capacity of a furnace is helpful in securing accurate control since it has the ability to level out any sudden thermal changes. However, in large annealing furnaces of high thermal capacity, there is a heat transfer lag due to the thermal resistance of the heating tubes and covers which is objectionable from the control standpoint. This condition results in a definite time lag between the temperature of the material and the controlling unit. If the thermocouple or other temperature-detecting device is placed at a considerable distance from the source of heat input, there is a period of time during which the heat is being applied that no effect is registered by the thermocouple on the controller. This so called "dead-time" or lag is quite detrimental to good control.

There are certain mechanical lags in the balancing and controlling mechanisms that are important when considering close temperature control. In mechanically balanced systems, a definite time interval is required for the pen movement to indicate the full change. However, with electronic units, this balancing period is quite rapid. Regardless of the type of instrument used, there is a definite amount of time consumed in the actual control operation of closing the fuel valve, either electrically or pneumatically. Where electrical-resistance heating is used, the time lag due to contactor operation is insignificant.

If the temperature lag is not too great and the thermocouple is properly located for maximum sensitivity, the simple on-off control may be used, particularly with electric furnaces. With this type of control equipment, the full power input yields a rapid heating rate which is ideal for quickly bringing the charge up to temperature, but may cause considerable cycling about the desired temperature. However, if the maximum heat input is not too large with respect to furnace capacity and selected temperature, good control can be secured with these modern controlling units.

In a large number of cases, the simple on-off type of control is not sufficient to do the job properly and for these requirements, the various types of automatic controllers were designed. These units are constructed to adjust the power input in proportion to the temperature difference between furnace and set point, and to properly compensate for continuous load changes. In certain types of control, the final uniform temperature of the furnace, determined by the action of the instrument, differs from the set point by a given amount of offset. A general example of this type is the multiposition, differential, gap-action controller which gives a definite neutral or no-action range between two fixed position (high and low or on-off). The control point may be located at any temperature within this neutral zone, depending upon the particular furnace load and the characteristics of the controller.

Under the action of the two-position type of controller, the furnace comes up to control temperature with maximum heat input, but instead of completely shutting off, a reduced heat input continues into the furnace. This reduced heat input may be manually set for a given furnace, load and temperature so that it will approximately balance the heat losses and minimize any excessive cycling effect. Used quite widely on gas- or oil-fired furnaces, this control can be actu-

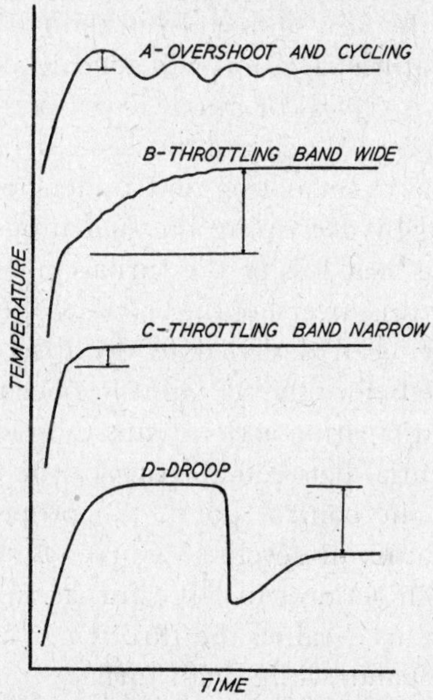

FIGURE 26. *Some Variations Encountered in Automatic Temperature Control*

ated by either or both the fuel and air valves. For electric heating, two individually operated sets of heating elements would be required for this type of control. By utilizing a third position, to shut all heat input off, a three-position controller is made which can be used effectively with batch-type heating where rapid initial heat input is desired and overshooting would be prevalent. The intermediate zone must be quite broad so that slight temperature fluctuations will not cause the heat input to shut off completely, as otherwise the control would not be much better than the on-off type.

If there is a heat transfer lag between the thermocouple and the heat source, a large furnace coming up to control point under full heat input will overshoot the desired temperature. When the thermocouple has reached the desired temperature, there is still a considerable heat differential between it and the heat source. Even though the fuel is shut off, this temperature variation will tend to equalize. Curve A in Figure 26 is typical of initial overshooting followed by a cycling condition. When the furnace reaches the required temperature, the overshooting and undershooting (cycling) may be reduced by decreasing the fuel input sufficiently to just balance the heat loss of the furnace.

To prevent this overshooting and subsequent cycling, controllers have been designed to decrease the heat input into the furnace before the set point has been reached. This throttling or anticipating action tends to gradually diminish the heat differential between all parts of the furnace so that, upon reaching the control point, the proper heat balance is secured for minimum cycling. Curves B and C in Figure 26 illustrate such action. In large furnaces, where the heat transfer lag is considerable, the throttling range is adjusted to cover a wide band, while small furnaces may only require a narrow band as shown by Curve C. Instead of turning the full heat input on or off at the control point, as in the simple on-off control, the throttling-type controller yields a band of action in which the heat input changes gradually from 100 percent on to 100 percent off. The throttling range may be so adjusted as to cover the full scale of the instrument or it may be narrowed down to on-off control.

The more throttling action on the furnace the more uniformly the charge comes up to temperature, but the rate of heating is greatly decreased, requiring longer cycles. However, in annealing operations, where good temperature uniformity is a necessity, no overshooting can be tolerated.

In certain types of control instruments, there is a fixed relationship between the control valve and the throttling

FIGURE 27. *Arrangement of Control Switches and Cams on the Leeds and Northrup Micromax Controller*

For a furnace having a continuous uniform load, this controller would be quite satisfactory. But an increase in load, requiring more heat input than can be supplied by the original valve-controller relationship, will cause the temperature of the furnace to drop and level off at a lower point. An mechanism. This means that for a given position of the throttling control unit, there is a fixed fuel-valve opening.

illustration of this "droop" or "offset" effect is given in Curve D of Figure 26. When the additional load is applied, the furnace temperature drops and then returns to a position lower than the original. The completely automatic controllers are so designed as to eliminate this droop characteristic.

The balancing mechanisms, as previously described, are solely a means of indicating the temperature of the hot junction of the thermocouple. However, in most applications, it is desirable not only to indicate temperature, but also to accurately control it. The simple on-off type control has been accomplished with several variations in mechanical arrangements, depending on the manufacturer. The main consideration is to connect, with maximum efficiency, the movement of balancing the slide wire to the actuation of a control switch or valve. For most instruments, a temperature-setting pointer must be added which can be manually adjusted to any desired temperature on the scale. Then, as the temperature rises to the set point, with the slide-wire contactor approaching the set-temperature indicator, the control switch is opened and the input is shut off. Figure 27 illustrates the Leeds and Northrup Micromax, which is used for securing close control.

The control contact switches are rotated to the desired position by means of the temperature-setting knob at the top front of the instrument. These switches are mounted on the large aluminum disk which is geared to the setting knob. The control temperature at which the switches are positioned is indicated by the pointer at the top of the temperature scale. By means of two notched fiber disks, mounted on the same shaft as the balancing slide wire, the contact switches are operated in the on-off control circuit. The position of the slide wire and of the control disks is translated into temperature by the lower pointer on the temperature scale. When both the set- and temperature-indicating pointers match, balance is secured with the on-off controls.

For the simple on-off control, the notches in both fiber disks are set in the same position.

With this particular instrument, as the furnace comes up to temperature, the front switch is open and the rear one is closed. When the desired temperature is reached, the notches in the fiber disks cause the special roller couplings to close the front switch and open the rear. When in this condition, the input power is off. At the balancing point, a slight clockwise rotation of the disks and slide wire will turn the power on, while a slight counterclockwise movement will shut it off. If the fiber disks are not positioned with their notches together, a differential control action is secured. By unlocking and rotating the front disk so that its notch

FIGURE 28. *Control Arrangement on the Brown Electronik Instrument (Courtesy, Brown Instrument Co.)*

is more advanced, the power input will continue after the rear switch is opened till the front switch is closed. As the furnace cools below control point the front contact will open first but the power will remain off until the temperature difference between the two fiber-disk settings is traveled. The rear contact then goes closed and the power turns on. The reverse condition occurs on heating: The rear contact opens and, after the differential is passed, the front contact closes and the power turns off. When the furnace is at the right temperature under differential control, the temperature-indicating pointer will vary over a range equivalent to the set difference between the on and off position. This differential type of control is beneficial in maintaining a uniform temperature of the material in cases where the control thermocouple is close to the heat input and the charge has a high heat capacity. The amount and position

FIGURE 29. *Mercury-Switch-Type Contactor on a Brown Electronik Instrument (Courtesy, Brown Instrument Co.)*

of this differential action can be adjusted to suit the particular furnace heating characteristics. Other fiber disks are switches may be added to secure additional control.

Another arrangement, as illustrated in Figure 28 (Brown Electronik), is actuated also by a fiber-cam movement. In this case, the cam C can be adjusted by loosening the knurled knob E. The temperature indicating pointer or pen is moved manually to the desired control point, then the notch in the cam is positioned so that the roller D will actuate the contacts A and B at this desired point. This adjustment is made at the back of the instrument, but a red pointer mounted on the front door can be moved separately to indicate the set control temperature. The small spur gear on the left is attached to the balancing mechanism and drives the large gear to which the control cam is fitted. On the same shaft is a drum around which a fine steel strand is wrapped. This strand continues to drive the balancing slide wire. The A-shaped gear sector in the background, driven from the same shaft, moves the recording pen for this circular-chart type instrument. A similar arrangement is shown in Figure 29 where the spring-contact switches are replaced by a mercury switch. This switch is adjusted by loosening the screw in the center of the fiber disk, then retightening, with the temperature indicator at the desired control point. This switch assembly is geared to the large spur gear which is connected to the balancing mechanism.

FIGURE 30. *Recording Pen and Control Movement of the Foxboro Potentiometer (Courtesy, Foxboro Co.)*

Another continuous-chart type instrument, Foxboro, with a straight-line slide wire, utilizes a different type of control unit. This control mechanism, illustrated in Figure 30, is actuated by the movement of the recorder pen. On the pen carriage is mounted the slide-wire contactor A and the con-

     Modern Pyrometry

trol-cam roller B. The wound slide wire runs parallel to the temperature scale of the instrument. The cam roller B contacts the edge of the curved, tapered section C, of stainless steel causing it to rotate. The axis of rotation of this curved section is parallel to the pen-carriage movement. Therefore, a horizontal movement of the pen carriage will yield an equivalent rotation of the curved section C. Section C runs the full length of the temperature scale and, at one end, its rotation is transferred to two fiber contact disks, (Figure 31). These contact disks and fingers can be adjusted so as to operate the control mechanism at any previously determined temperature.

FIGURE 31.   *End View of the Control Mechanism in Figure 30 (Courtesy, Foxboro Co.)*

Where simple on-off control is required, the small compact control unit of Foxboro can be used. This unit is illustrated in Figure 32 and operates on the potentiometer principle. With this means of control, the desired temperature is set by a large graduated dial, which actuates the slide wire. This fixed slide-wire position then produces a defi-

FIGURE 32.   *General View of the Foxboro Controller (Courtesy, Foxboro Co.)*

nite amount of millivoltage from the battery circuit. The millivoltage of the hot junction of the thermocouple must then balance the preset amount of millivoltage in order to bring the galvanometer to null position. When the temperature is below the set point, the galvanometer will deflect to the low side as illustrated in the lower section B of Figure 33. In this instrument, the sensing mechanism is a flat detector bar which is depressed approximately every 10 seconds. The bar hits the top portion of the galvanometer needle and, in this condition, the mercury switches or relays are actuated to increase the fuel supply. In the upper section A, the set temperature has been reached and the galvanometer is in balance position. The depressor bar is unobstructed and will cut off or reduce the heat input to a minimum. Automatic cold-junction compensation is obtained electrically by inserting a small calibrated coil of wire in the circuit which changes its resistance with any changes in the surrounding temperature. It is possible to operate a large number of these instruments side by side on a panel, utilizing one interconnected driving mechanism and a multipoint recorder for the group, thus making a relatively compact, inexpensive control and recording station.

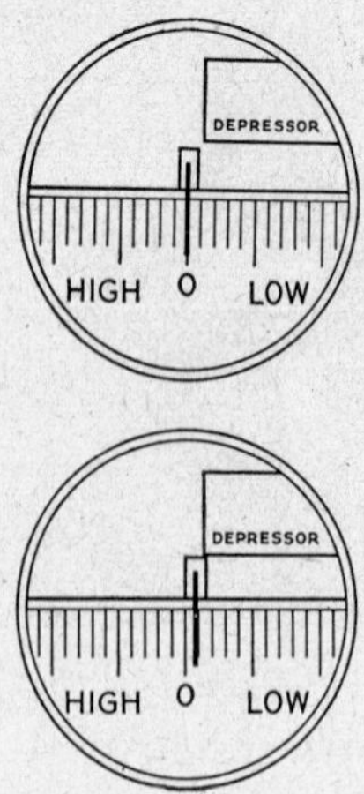

FIGURE 33. *Relationship between the Depressor Bar and Galvanometer Needle in the Foxboro Controller (Courtesy, Foxboro Co.)*

The photoelectric-cell type Tagliblue instrument utilizes another control arrangement. Off-and-on spring-contact switches are located above the pen carriage. They are behind, but attached directly to, the temperature-setting indicator. The temperature-setting knob will move this switch assembly and indicator to any desired position on the temperature scale. A flexible lead wire is attached to these contacts which permits accurate settings of this control unit. A small gear segment is firmly attached to the top of the pen carriage. A movable combination fiber gear and cam segment is mounted on the control unit above the pen carriage. A slight rotation of the fiber gear segment causes the upper cam portion to actuate the control switches. As the temperature increases, the gear-teeth segment mounted on the pen carriage approaches and then engages the movable gear segment on the control unit and the on-off control goes into effect to open the power circuit.

FIGURE 34.   *Activating Mechanism of the Bristol Electronic Controller (Courtesy, Bristol Co.)*

The next instrument utilizes a somewhat different technique from the previously described types. Figure 34 is the side view of the indicating and controlling mechanism of the Bristol Electronic Controller. The indicating mechanism is a cold-junction-compensated millivoltmeter, similar to those previously described. This instrument is converted to an electronic controller in the following manner. On the temperature-setting arm are two small flat wound coils positioned so as to face each other, but with a small clearance between them. These coils are connected in a high frequency oscillating circuit which also actuates the control relay. The millivoltmeter pointer carries a small, light-weight metal flag which is positioned so as to pass between the coils mounted on the temperature-setting arm. As the temperature of the thermocouple hot junction rises and the millivoltmeter pointer approaches the control point, the flag passes between the two coils, thus changing the characteristics of the oscillatory circuit. This movement will decrease the current output from the electron tube and, by means of a relay, control the fuel input.

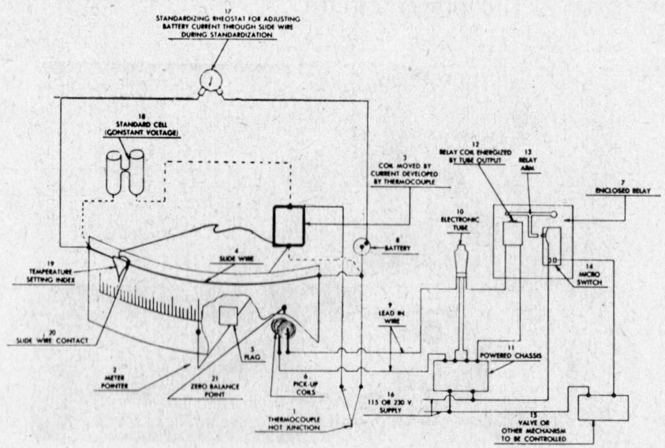

FIGURE 35.   *Schematic Diagram of the Wheelco Electronic Potentiometer-Type Controller (Courtesy, Wheelco Instruments Co.)*

The Wheelco Instruments Company has a similar electronic coupling in a potentiometer-equipped indicating controller, Figure 35. The movable temperature-setting arm is in contact with the slide wire of a potentiometer circuit. The two high-frequency oscillator coils are permanently at the zero point of the null-galvanometer. The position of the temperature-setting arm will determine the amount of millivolt output from the battery circuit which the thermocouple must balance when the desired temperature is reached. After the temperature-setting indicator has been moved to the required scale temperature, the galvanometer indicating pointer is considerably out of balance, but it approaches the null point as the furnace temperature rises. A thin aluminum vane, mounted on the galvanometer arm, passes through the two coils when the desired temperature is reached and then the electronic control goes into effect.

A recent innovation by the Wheelco Instrument Company has been an arrangement which utilizes this particular electronic coupling action to produce a recording controller. The change consists of connecting the oscillator pickup coils in a circuit to operate a balancing motor instead of control relays. This control system is composed of snap switches and control cams. In the case of the direct-deflection type of instrument, the aluminum vane is attached to the millivolt-meter arm, the same as before. The oscillator coils are mounted on a follow-up arm which, in turn, is fastened to a vertical shaft geared to the balancing motor. This motor is geared also to a horizontal spiral shaft which drives the recording pen on the chart. The chart itself is driven by a separate synchronous motor, with an additional motor (shaded-pole induction type) driving the rewind to assure a taut chart at all times.

As the millivoltmeter arm, with the attached vane, moves upscale during the heating cycle of a furnace, the balancing motor drives the follow-up arm upscale until the two mounted oscillator coils are in balance position with respect to the vane. This balancing motor is energized when the metal vane is not between the oscillator coils. Thus any movement of the vane is closely followed by the motor-driven arm, the direction of which is determined by the position of the aluminum vane between the oscillator coils. A mechanical stop is positioned so as to prevent any vane movement from these coils on the upscale side. This prevents a false direction of motor rotation.

Utilizing this method in the null-type measuring system, the two oscillator coils are on a stationary mount, with the aluminum vane located on the galvanometer needle. Thus any movement of the vane between the coils will regulate the amount of rotation and direction of the balancing motor. This, in turn, moves the slide-wire contact to balance the potentiometer circuit. The same balancing action drives the recording pen as previously described.

The action of this electronic system may be more clearly described as follows. The oscillator pickup circuit actuates the proper power-supply tube in the motor-drive circuit to move the reversible motor in the proper direction for balancing. When the galvanometer arm is below control point, the vane is away from the oscillator coils. This permits complete interaction between the two coils and a current flows in the pickup circuit. The current affects one of the two triode tubes in the power circuit, which produces the proper phased power to run the motor in the upscale direction. At the point of balance, the vane on the galvanometer arm is partially between the pickup coils. This reduces the current flowing in the pickup circuit, thus decreasing its effect on the power tube so that the output of one power tube balances that of the other and the balancing motor is held stationary. However, as soon as the vane moves farther between the coils, the current is decreased still more so that the opposite triode power tube overbalances the first tube to drive the balancing motor downscale.

The control system of this instrument has been changed from relays actuated by the oscillator coils to cams and snap switches. On the vertical follow-up shaft, running from the measuring system to the motor, are mounted the control cams. These are cut so that a small rotation at a given point on the circumference will cause a roller snap switch to open or close rapidly. The control switches can be positioned to any desired temperature-control point by adjusting a control-setting index pointer on the temperature scale. The index arm is connected to a hollow vertical shaft (slipped over the vertical follow-up shaft) to which is attached a disk holding the roller snap control switches. The control index thus will position the snap switches and, as the furnace comes up to the desired temperature, the balancing motor will rotate the control cam to the cut portion where the on-off control will go into effect.

The modern proportioning controlling unit is relatively

complicated as compared to the simple on-and-off control. These new units are designed to adjust the power input in proportion to the temperature difference between furnace and set point and to properly compensate for continuous load changes. In other words, as the furnace comes up to the set temperature, full-power input is employed so that heating is rapid. But, in order to prevent the furnace temperature from overshooting the set point, this proportioning controller begins to decrease the power input to the furnace before the desired temperature is reached. This enables all parts of the furnace and its charge, to arrive uniformly at the desired temperature. The width of this proportioning range can be adjusted for the requirements of the furnace.

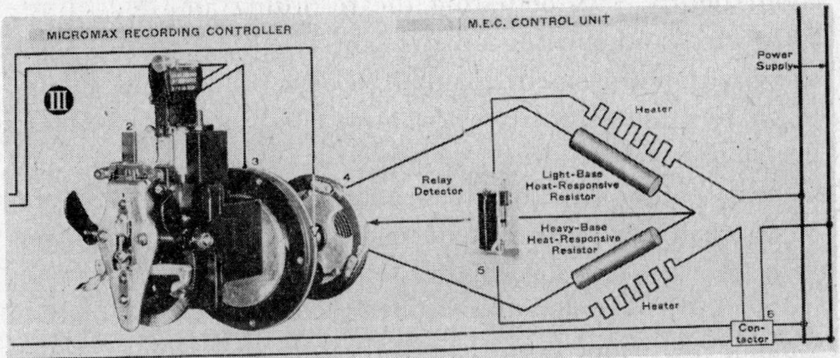

FIGURE 36.   *Schematic Diagram of the Leeds and Northrup Micromax Electric Control Unit; Duration-Adjusting Type (Courtesy, Leeds and Northrup Co.)*

Figure 36 illustrates the Leeds and Northrup Micromax Electric Control unit attached to the recording controller. Attached to the measuring slide wire of this unit is the slide wire for the bridge-type control circuit. In this circuit, there are two heaters connected to the power supply, which affect two heat-responsive resistors. One resistor has a heavier base (more heat capacity) than the other, so that it heats up more slowly. The two heat responsive resistors act as anticipators to any furnace-temperature change, since they are much more temperature sensitive than the furnace. The unbalance in the control bridge is detected by means of a relay detector 5 which energizes the contactor 6 in the power-supply line. When the furnace is approaching a set temperature, the Micromax recording controller moves close to the control point and, at the same time, the controlling slide wire 4, connected to the recording instrument, begins to balance the control bridge. Both the heavy-base and light-base resistors are heated by individual heating units connected to the power supply. With both resistors hot, the control circuit is balanced by movement of the slide wire 4. After the set temperature is reached, the control circuit deenergizes the contactor, allowing the resistors to cool. The furnace then cools slightly, thus actuating the recording and controlling slide wires and turning on the power. However, this time, the power requirement is small because the furnace is close to the set temperature. Now, the effect of these resistors becomes apparent since the power is on long enough just to heat the one light-base resistor. The power requirement being small, it is not sufficient to affect appreciably the heavy-base resistor. This heating of the light-base resistor increases its electrical resistance in the bridge circuit which tends to balance the bridge even though the furnace has not reached the set point. This balance is accomplished with very little movement of the control slide wire. Once this balance is secured, the contactor opens and the light-base resistor cools quickly. The electrical resistance of this portion of the bridge drops and unbalances the circuit, thus turning the power on again. The separate balancing action due to heating and cooling this one resistor takes place with practically no movement of the recording or controlling slide wires and the power input thus secured is sufficient to maintain the furnace at control point. To prevent temperature droop due to sustained load changes, the heavy-base, heat-responsive resistor is a necessary part of the circuit. As the temperature of the furnace decreases due to increased load, the recording controller follows this drop, unbalancing the control circuit and energizing the contactor. As the furnace is heated, both heat-responsive resistors are also heated. But, since the temperature drop was considerable, the heavy-base resistor will heat up and prevent the light-base resistor from completely controlling the input. Consequently, the balance in this case must be attained by the control slide wire 4 and a new temperature input balance must be achieved. Additional flexibility is obtained by a series of variable resistances which can be set to maintain a desired throttling range, proper rate of droop correction and the action-rate sensitivity of the contactor.

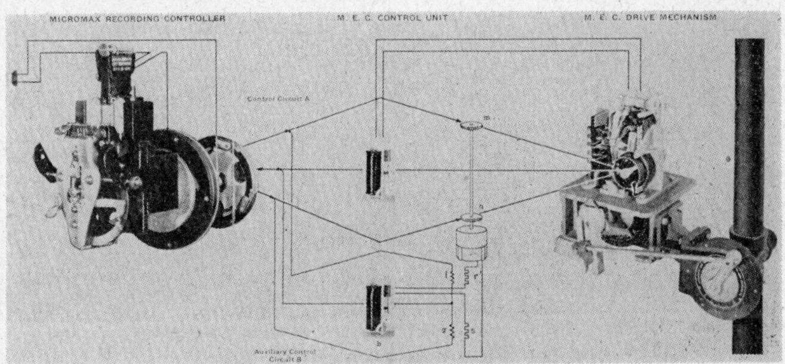

FIGURE 37.   *Schematic Diagram of the Leeds and Northrup Proportioning Unit for Gas or Oil; Electric-Control Type (Courtesy, Leeds and Northrup Co.)*

For valve-operated gas or oil equipment, an arrangement similar to that shown in Figure 37 may be used. This system is composed of four sections: Micromax recording controller, control circuit A, auxiliary control circuit B, and the valve-driving mechanism. In general, the system operates in the following manner. When the thermocouple detects a change in furnace temperature, the Micromax instrument rotates the measuring slide wire; the control slide wire, being on the same shaft, is rotated as well. This rotation will unbalance the entire control system. In the unbalanced auxiliary circuit, the relay *b* energizes the droop-corrector motor, which turns the rheostats *m* and *n* in circuit A. The relative position of the new recorded furnace temperature with respect to the set point is the factor determining whether the resistance of *m* and *n* increases or decreases. This furnace temperature—set point relation also determines which of the heaters *r* or *s* is energized. These heaters affect the temperature, and thus the resistance, of coil *f* or *g* respectively. Increasing the resistance of *f* or *g* will tend to balance circuit B and stop the droop-correcting motor. However, as soon as this occurs the heater *r* or *s* cools causing a decrease in resistance of *f* or *g*, thus unbalancing the auxiliary circuit again. This action continues in circuit B until resistance coils *f* and *g* are equal at the control point. As this action occurs in the B circuit, balance is being secured by the slide wire of the valve mechanism in circuit A. A motor drives both the balancing slide wire and the valve. Thus, as the motor changes the fuel-valve position the attached slide wire brings the control circuit back into balance. The unbalance in the control circuit A actuates the relay which energizes the valve-drive motor. Without the droop-corrector influence of rheostats *m* and *n*, there would be a fixed relationship between the control slide wire and the slide wire on the valve-drive motor. An increase in the furnace load would require a new valve position to maintain the desired temperature. The auxiliary circuit B adjusts the rate of droop-correction through the action of the heater units on the droop-correcting motor.

If explosive atmospheres may be encountered, it is desirable to operate the fuel valve with something other than an electric valve motor and slide wire. For this application, a pneumatic control can be used. In pneumatic systems, completely automatic control is secured by the use of three types of action: throttling, proportioning, and reset (droop-correction). The throttling range (proportional band) is manually set for the specific control process. When properly adjusted, the throttling range is sufficiently broad to prevent the temperature from overshooting the set point during heating. This setting determines the temperature at which the controlling mechanism goes into action. A narrow proportioning range will result in a large amount of valve motion from a small temperature deviation. The proportional control is an opposing or corrective action which produces a fixed relationship between the furnace temperature and fuel-valve position. That is, when the furnace temperature

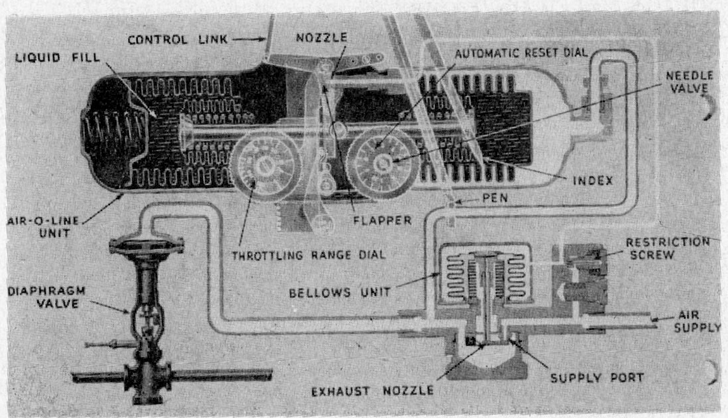

FIGURE 38.   *Operating Mechanism of the Brown Pneumatic Controller (Courtesy, Brown Instrument Co.)*

decreases, the fuel valve is opened an equivalent amount. If the furnace load is changed, however, a new relationship must be established. This is accomplished by the reset action, whose rate is governed by a restricted pneumatic or hydraulic flow. This rate is manually adjusted to yield proper offset correction for the given process.

The Brown air-operated electronic instrument, illustrated in Figure 38, is of the circular-chart type. This unit is small and fits into the regular instrument case, with the controlling and recording mechanism. As illustrated, a control link connects a flapper-actuating rocker arm to the shaft which positions the pen. Thus, any pen movement is reflected by a tilting of the rocker arm. By connecting the control link to the opposite end of the rocker arm, reverse action is secured. The set-point index also positions the rocker arm through a differential lever. Therefore, after the initial positioning by the index, the rocker arm moves simultaneously with the pen. Just below the rocker arm is a flat, vertical, spring flapper which covers the nozzle opening of the small air line. The rocker arm is designed so that, by a movement about the center-bearing, contact can be made with the end of the flapper to move it back from the nozzle of the air line. An indication of the action of this unit can be secured best by examining the movements during the control of the temperature of a furnace. The rocker arm is tilted away from the flapper by the temperature-setting index. As the furnace temperature increases, the pen moves upscale and rotates the control rocker arm back to contact the flapper. At the beginning of the throttling range, the rocker arm contacts the flapper and moves it back from the air nozzle. This produces a pressure drop in the air line, which, amplified by a pilot valve, lowers the pressure on the control-valve diaphram, thus decreasing the amount of fuel going into the furnace. This pressure drop is transmitted from the pilot valve to the right side of the main bellows of the control assembly. As a result, this large bellows, inner bellows and inner connecting arm move to the right. The movement of the connecting arm readjusts the flapper against the nozzle sufficiently to stabilize the pressure of that line. If the furnace is to be brought up to the control temperature, the pressure on the control valve must continue to decrease a small amount. This condition is obtained through the automatic reset—a needle valve connecting the liquid fill between the right and left dual bellows combination. As the initial pressure decrease in the right-hand bellows pulls the inner connecting arm to the right for proper flapper-

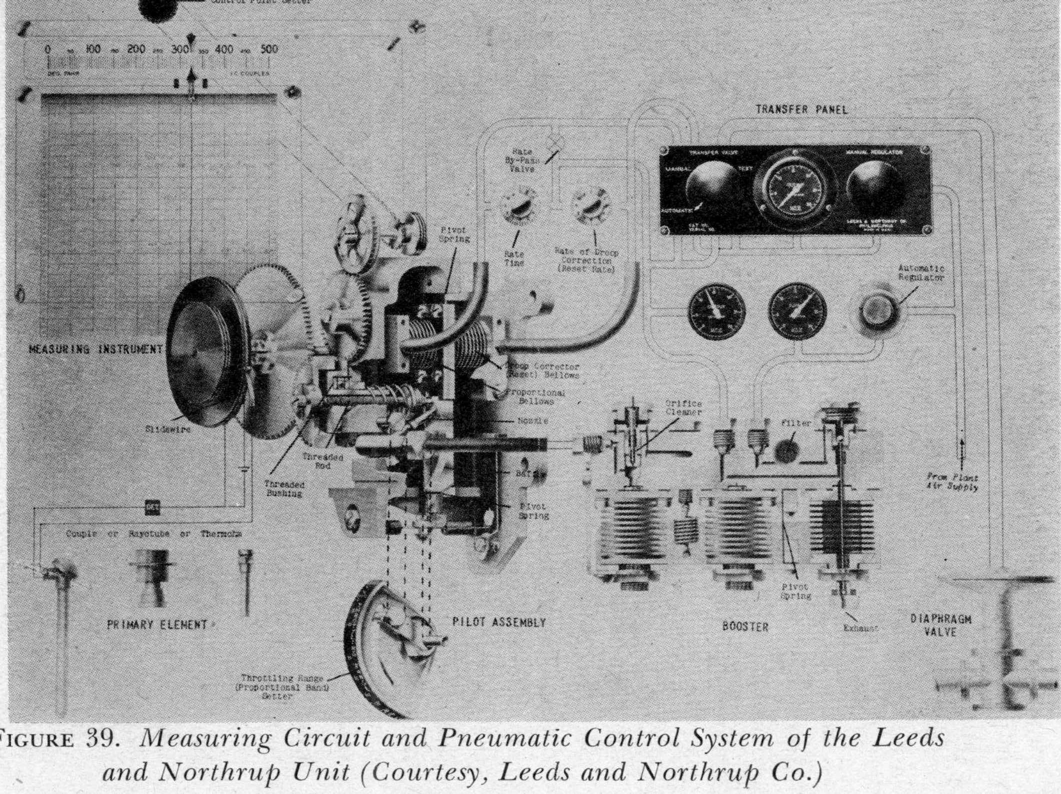

FIGURE 39. *Measuring Circuit and Pneumatic Control System of the Leeds and Northrup Unit (Courtesy, Leeds and Northrup Co.)*

nozzle adjustment, the spring tension on the left-hand bellows increases. This liquid pressure is equalized by flow from left to right through the small, automatic reset needle valve. As this condition occurs and the small spring-loaded inner bellows return to their normal position, the flapper is pulled away from the nozzle to decrease the pressure once more. This proportioning and reset cycle is continued until the set-point temperature is reached.

The pilot valve in this diagram amplifies small pressure changes in the nozzle line by the combination of two bellows, the low nozzle pressure acting on the large bellows to balance the high pressure on the small bellows. Any pressure differential operates a small flapper, covering both the supply and exhaust ports. If the total pressure on the inside of the small bellows is greater than that on the outside of the large bellows (the furnace is above the set temperature) the inner bellows and its hollow stem move upward. This permits air from the main air line to flow past the flapper into the hollow central stem, exhaust into the area between the two bellows and escape. Since this produces a pressure decrease in the main line, the return spring on the fuel valve tends to balance the opposing pressure by partially closing the valve. When opposite conditions occur, the large bellows pushes down the small bellows to open the main air-pressure line, thus increasing the pressure on the fuel valve sufficiently to open it.

Another type of pneumatic balance system, Leeds and Northrup, utilizes three opposed-pressure bellows acting on a balance arm. Figure 39 is a general illustration of the action of this system. The measuring circuit is electrically balanced as previously described in the operation of the Micromax instrument. The control-point-setting knob on the front of the instrument will rotate a threaded bushing through which the threaded control rod travels. Rotation of this bushing on the threaded rod produces a forward or backward motion of the latter. The shaft of the measuring slide wire is directly connected to the horizontal control rod, so that any motion of the measuring system will be directly translated to motion of the threaded control rod. During the control action, the slide-wire movement rotates the rod while the bushing remains stationary. Thus, the position of the threaded rod is moved in two ways. The positioning and subsequent controlling movement of this rod is translated, by contact, to the vertical, curved baffle section. This contact is made at the top of the baffle, thus producing a forward or backward tilting of the curved section about a spring pivot at its bottom.

The air pressure in the system is fed from a regulated, restricted supply to the exhaust control nozzle. This nozzle is positioned so that its exhaust is controlled by the proximity of the curved baffle section. When the baffle is moved toward the nozzle, the air pressure inside the system is raised, and when tilted away from the nozzle, air is permitted to exhaust, thus lowering the pressure of the system.

The proportioning action and reset-rate action of this instrument is secured by a vertical arm, connected to the control baffle, which operates between two opposed bellows. If the threaded control rod moves back and allows the baffle to move toward the nozzle, the pressure of the system will increase. This pressure increase acts on the proportioning bellows which opposes the movement of the control rod by moving the baffle away from the nozzle. This action decreases the pressure of the system. Such motion requires a relatively large travel of the threaded control rod to bring about the full change in output pressure. An adjustable valve in the line makes it possible to regulate the rate of the proportioning action. To regulate the width of the proportioning band, the nozzle itself is turned either up or down, the baffle having the proper contour to make contact with the nozzle on any possible position. When the nozzle is in the uppermost position, its contact with the baffle is at the point of maximum baffle movement since it is the end away from the pivot. This increases the effect of the recorder motion as compared with the opposing effect of the proportioning bellows, thus yielding a narrow proportioning band. By lowering the nozzle, the actual operating movement of the baffle is less, being closer to the pivot point. In this case, the effect of the recorder action is quite small while the opposition of the proportioning bellows is quite substantial. This condition yields a wider proportioning band. To provide the proper reset action for preventing a temperature droop due to a load change in the furnace, another bellows is placed so as to oppose the action of the proportioning bellows. The reset valve being partially open will permit the pressure of the line to slowly pass through and affect the reset bellows to balance the proportioning bellows. By proper setting of the rate valve, the correct degree

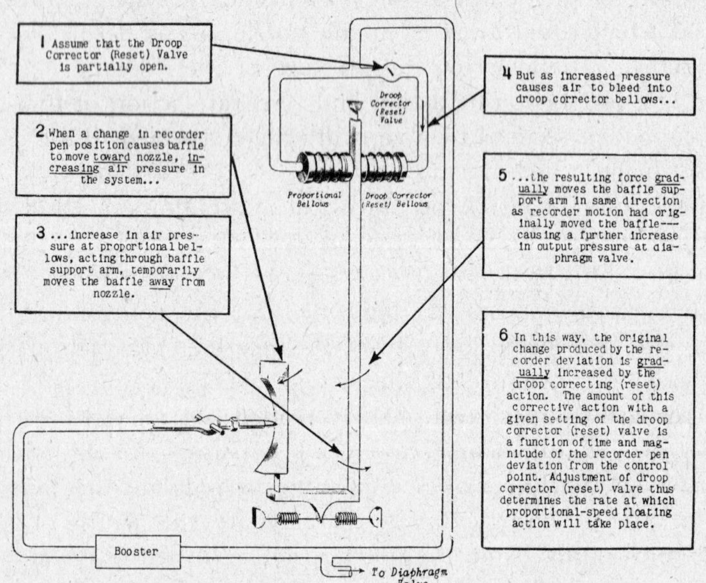

FIGURE 40. *Method of Securing the Correcting Action in the Leeds and Northrup Pneumatic Controller (Courtesy, Leeds and Northrup Co.)*

of proportioning can be obtained for a given furnace to prevent cycling. The reset is adjusted by its rate valve to take care of any sudden or sustained load changes of the furnace. Perhaps a clearer idea of this action can be secured from the sketch of Figure 40.

Since only a small air flow passes through the control nozzle, a booster element is required to effectively operate the diaphram valve. As shown, the nozzle pressure is operative on the first bellows unit. The two bellows units to the right of the first operate on the full line pressure. The first bellows, although operating on relatively low pressure changes, is sufficiently magnified by means of the lever system to control the full line pressure. When the first bellows has a pressure increase, the resultant motion of the booster system is a slight counterclockwise rotation of the bottom lever arm about the pivot spring, which opens the inlet ball valve to increase the pressure on the diaphram valve. A decrease in pressure on the first bellows results in the operation of the exhaust ball valve and a reduced pressure on the diaphram valve.

It might be well to follow this action through again from a temperature change in the furnace as detected by a thermocouple. As the temperature rises above the set point, the slide wire of the recording instrument is rotated to balance the additional millivolt output of the thermocouple. Such slide wire rotation moves the threaded control rod away from the baffle, permitting the spring action to move the baffle toward the nozzle and increase its air pressure. This pressure increase is reflected to the bellows booster system which increases the air pressure on the diaphram valve and decreases the amount of gas to the furnace.

Another pneumatic control system, illustrated in Figure 41, is the Foxboro Stabilog. This instrument is shown as controlling pressure in a vessel. However, a similar pneumatic control system is utilized in the control of temperature by substituting a temperature-measuring unit previously de-

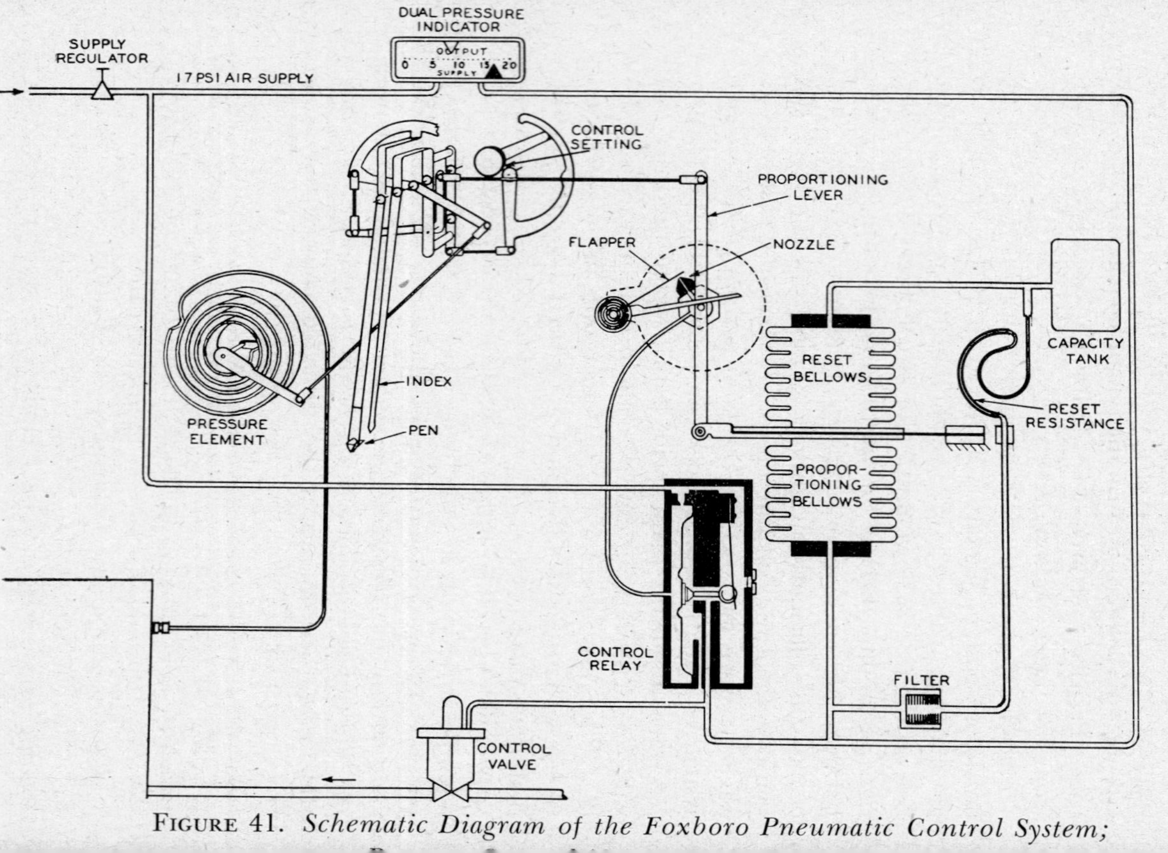

Figure 41. *Schematic Diagram of the Foxboro Pneumatic Control System;*

scribed. In this unit, only two opposed bellows are used, one for proportioning and the other for reset action. Instead of having valves to control the rate of action of these bellows, a flattened bourdon tube (one end fixed and the other end adjustable) is used which can be positioned to give any desired air-flow resistance. A bourdon tube has an elliptical cross section and is bent into a circular form. One end is connected to the rate-setting index while the other is in a fixed position. Uncoiling this tube, by turning the rate index, changes its cross section from elliptical to circular, thus decreasing its resistance to air flow. The bellows in this system act on an arm connected to a proportioning lever arrangement. This proportioning lever is designed so that the operation of the two opposed bellows will yield a vertical motion while the differential linkage of the indicating mechanism will produce a rotational movement about its connection to the horizontal bellows arm. A pin on this proportioning lever engages the contact arm which, in turn, positions the flapper with respect to the nozzle. The differential linkage between the pen arm and index pointer is arranged so that any motion of the pen relative to the index pointer results in a motion of the flapper, and when the pen arm and index pointer are together, at set point, the flapper becomes tangent to the nozzle, thus permitting no drop in air pressure in the system.

The width of the proportioning band is changed by rotating the flapper contact arm, flapper and nozzle simultaneously about the proportioning lever pin. As the flapper contact arm is rotated to a vertical position, there is maximum motion of the flapper for a given pen motion and no motion from any bellows change. This yields an on-off action. From this zero proportioning band we go through the full range to an infinite proportioning band when the flapper contact arm assembly is moved to the horizontal position. Here the pen movement yields no flapper motion, with a small bellows change giving a large flapper action. If the flapper is rotated still further, the proportioning band again becomes narrower and the controller action will be reversed; that is, a drop in the pen arm with respect to the index will cause a decrease in pressure on the control valve instead of an increase.

The variations in control pressure at the nozzle are accentuated at the control relay to operate the main air supply for the control valve. A build-up in nozzle pressure will increase the pressure on the diaphram in the control relay, thus forcing open the small ball valve on one side and simultaneously closing the other side with the small conical seat. This small cone seals off the exhaust, while the opening of the ball valve permits the line pressure to flow through and down the tube to the control valve. If the nozzle pressure decreases, the diaphram pressure in the control relay also decreases, thus allowing some of the air from the control-valve line to be exhausted, since the ball valve closes and the cone valve opens to the exhaust.

In the case of the Bristol pneumatic instrument, the temperature-measuring system is connected through differential linkage to the arm of the free vane. This flat metal vane is free to rotate between two control nozzles located about 0.007 inch apart. The air escapes from these nozzles when the vane is withdrawn, thus decreasing the pressure on the control system. When the temperature-measuring mechanism rotates the vane to a position between the nozzles, the result is a pressure increase that is reflected to the pilot valve and from there to the control valve. The diaphram-operated pilot valve is somewhat similar to that in the Foxboro system. Operating from the controlled line pressure are the reset and proportioning bellows which are connected to the free vane in an opposite direction to that of the temperature-measuring mechanism. The degree of reaction of these bellows is adjusted by means of valves.

When pneumatic control is used, a constant pressure source of clean, dry compressed air is required to obtain good, trouble-free action. These instruments are designed to compensate for small changes in line pressure. Various types of operating valves are used, depending on the requirements of the system, ranging from the spring-opposed diaphram type to the large pressure-cylinder-action port valves. These various types of valves are suitable for control of steam, high-pressure liquids, gas and oil.

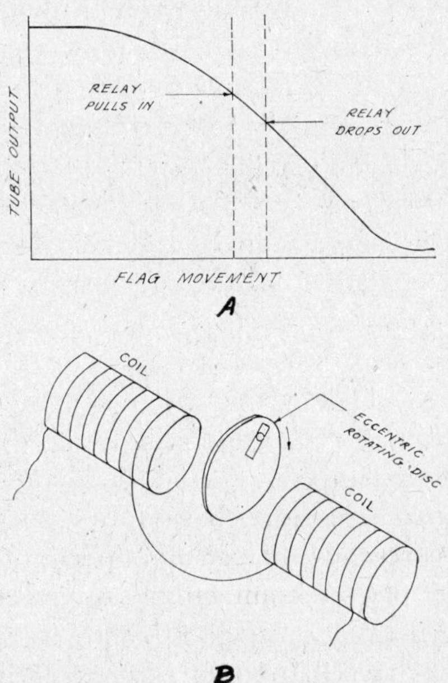

FIGURE 42. *A. The Effect of Flag Movement in the Electronic Millivoltmeter-Type Instrument. B. The Rotating Proportioning Disc (Courtesy, Wheelco Instruments Co.)*

In the case of the electronic millivoltmeter type of instrument, a method of proportioning control, as developed by Wheelco, is clearly illustrated by the diagram in Figure 42. The instrument utilizing this type of control has a metal flag on the galvanometer needle which passes between high-frequency oscillator coils at the set point. As the furnace approaches the control point, the flag moves into the coil

control system. Such shielding decreases the effective power output from the tube circuit (illustrated by the upper portion of Figure 42) to the point where the relay becomes de-energized and the contactor is opened. When the furnace cools slightly, the flag moves out of the coil system, thus increasing the tube output and energizing the relay. A flag movement of approximately 0.006 inch is sufficient for control action. The proportioning effect is obtained by inserting a device, the nature of which is illustrated in the lower portion of Figure 42. This unit is placed in the oscillating electrical circuit in series with the coils acted on by the galvanometer control flag. When slowly rotating the eccentric cam between this separate set of coils, the effect is the same as if the control flag had moved back and forth between its set of coils. When the cam in this auxiliary unit passes between its coils, the net effect is to drop the tube output of the circuit and as the cam moves out, the tube output returns to normal. This cycling effect will decrease the tube output for each half revolution of the cam. As the furnace approaches the set temperature and the galvanometer needle, with the flag attached, moves close to the oscillating control coils, a point is reached, below the set point, where the unbalancing effect of these two actions is sufficient to drop the tube output to a value that will allow the relay to drop out. However, the relay will pull in as the rotating cam moves out of its coils. This produces an anticipating action which not only proportions the input to the furnace at the set temperature, but also prevents initial overshooting. Three speeds of rotation of this proportioning cam can be secured; 15, 30 or 60 seconds per revolution. The cam may be adjusted to various degrees of eccentricity, thus changing the degree of pulsation and the width of the proportioning band.

The power input may be proportionately controlled by the use of additional equipment connected to a regular on-off controller. Figure 43 is a schematic illustration of an instrument of this type. This unit automatically determines the correct amount of full power "on time" that is necessary to reduce under and overshooting a given furnace temperature. Riding on a specially designed rotating cam is a

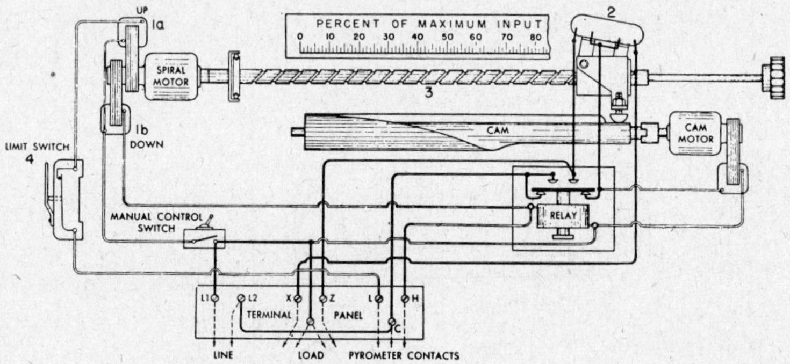

FIGURE 43.   *Auxiliary Input-Control Unit (Courtesy, Wheel-co Instruments Co.)*

mounted mercury switch which regulates the power input into the furnace. On the cut-away section of the cam, the rider allows the mercury switch to remain in the "on" position whereas the built-up cam section raises the mercury switch rider, thus shutting the power off. The cam is so designed that its cut away portion ranges from 100 percent to 0 percent of the circumference in a continuous form. Therefore, the power can be "on" 100 percent of the time at one end of the cam or "off" 100 percent at the other end, with any percent "on" time possible between the two extremes. The position of the mercury switch with respect to the rotating cam is regulated, using a spiral grooved rod driven by a reversible motor. The downscale motor is actuated as soon as the furnace temperature goes above the set point on the main control instrument and the mercury switch moves to the left. This downward movement continues until the furnace temperature falls below the set point. At this moment, the driving motor is reversed moving the mercury switch toward increased input time until the set point is reached and the entire cycle is again repeated. This action is continued until the proper balance is secured. The percentage of input time is, therefore, brought to the desired value by the relationship of the time above set point to that below set point.

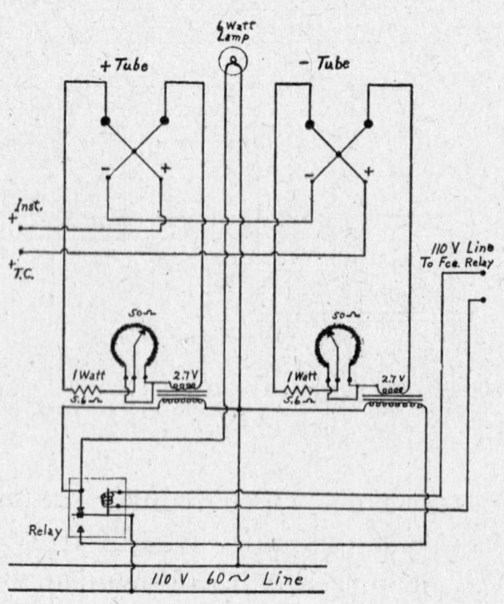

FIGURE 44.　*Wiring Diagram of the Xactline Controller Unit (Courtesy, Claud S. Gordon Co.)*

Another ingenious device, designed to yield straight-line control and to prevent temperature overshooting, has been put on the market by Claud S. Gordon Co. and is called the Xactline. Figure 44 shows a diagram of this instrument. This anticipator utilizes two small individually glass-sealed thermocouples with a small heating element adjacent to each. One small thermocouple is connected so that its millivolt output is additive to the furnace thermocouple while the output of its neighbor is connected in a subtractive manner. When the furnace control is "on," the heating unit adjacent to the additive thermocouple is heated. The millivoltage of this heated thermocouple, being added to the furnace thermocouple, is sufficient to actuate the control before the furnace temperature reaches the set value. This prevents the furnace from overshooting its set temperature. As the furnace power is turned off, the heater unit of the additive couple also goes off, diminishing the effect of this external source of millivoltage. The heating element located beside the subtractive thermocouple unit is energized as soon as the furnace power is turned off. This is accomplished by means of a relay in the system which closes this small thermocouple heating circuit when the furnace power goes off. The output of this couple is connected negatively to the furnace couple, permitting the controlling unit to turn the furnace power on a little before the furnace couple itself would indicate. Straight-line control is obtained by the proper balance of these two small opposing outputs. By the use of a small rheostat, connected to each of the small heating elements, the millivoltage output of each small thermocouple can be varied to suit the operating conditions of the furnace.

Another type of controller, which increases the sensitivity of the thermocouple, has been designed by Westinghouse Electric Co. Two small thermocouple elements, along with a heating element, are sealed in an evacuated glass envelope. The main difference between the two small thermocouples is their thermal capacity. These two couples are connected in series to the furnace couple in such a manner that the small control couple with low thermal capacity is additive while the other control couple with high thermal capacity is subtractive. These fine control couples are equidistant from the small heater element, which is energized by the power contactors of the furnace. When the contactor is "in," the heater element in this unit is heated as well as the furnace itself. The thermocouple element with the low thermal capacity heats up more rapidly than its partner, yielding an over-all additive effect to the furnace couple. This added increment will cause the furnace control to operate before the actual furnace temperature has reached the set point. When the controller acts to pull out the contactor, the small heating element in this anticipatory unit cools as do the two small thermocouples. The thermocouple with the largest thermal capacity retains its heat longer and thus the over-all effect of this unit becomes subtractive. This condition causes the furnace controller to actuate the power contactor before the temperature of the furnace has reached

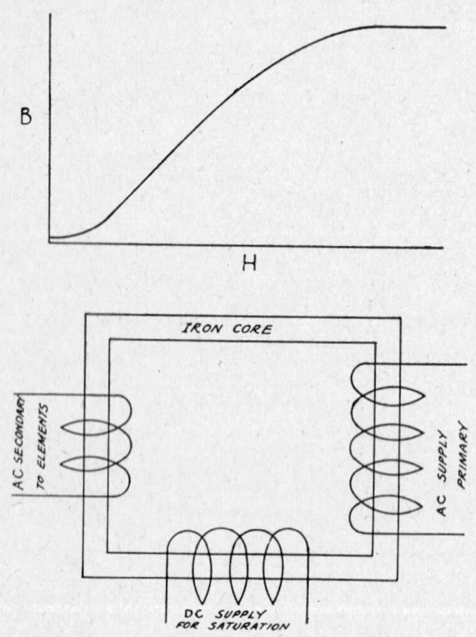

FIGURE 45. *The Operating B-H Curve and the Control Reactor Used in Proportionally Controlling Electric-Power Input*

the set point. However, during the initial heating cycle, the heating element in the high-capacity unit is energized sufficiently to bring both of the fine-control thermocouples to the same temperature, cancelling their effect. This would permit a certain amount of overshooting as the furnace first reaches its set point.

For most electric-furnace applications, the pyrometer-actuated controls are of the on-off type. That is, when the temperature goes too high, the power is off and when it is too low, the full power is turned on. Close control is obtained only through relatively high-speed contactor action, operating on the full-power load. Another method of electrical-power input control has been designed to eliminate contactor action and yield a stepless range of input. In this arrangement, the incoming power line is connected to one winding of an iron-core reactor (transformer), while the winding on the other outside leg is connected directly to the heating elements. The center leg of this reactor has a winding to carry a direct-current supply controlled by thyratron tubes which, in turn, are actuated by the controller. A simplified wiring diagram of this reactor is shown in Figure 45. The upper portion of the figure illustrates the magnetization curve for iron. The magnetizing force, H, is plotted on the abscissa while the lines of flux, B, are plotted on the ordinate. As the magnetizing force H is increased, the flux B increases and reaches a maximum (saturation) at approximately 21,000 gausses. Further increase of H will yield only a negligible increase in B.

As the temperature approaches the control point, the controller operating the thyratron tubes permits more direct-current power to flow into the center winding. This increase in direct current raises the magnetic flux in the iron core toward the saturation value. When the iron core approaches saturation value as a result of the direct-current flux, the amount of unsaturated flux available for reversal by the incoming alternating-current power winding approaches zero. When the reactor is completely saturated with direct-current flux, the incoming alternating-current power produces no flux reversal in the iron core. As in any transformer, it is the amount of magnetic-flux reversal that induces a voltage in the secondary winding. Since the alternating-current flux is reduced, the induced voltage in the secondary winding is zero and no power is supplied to the heating elements. If the temperature is below the control point, the amount of direct-current power on the reactor is decreased. This produces a drop in the direct-current flux, allowing more alternating current flux to flow in the reactor and to induce more power in the element winding. In this manner, the input power can be carefully proportioned to produce straight-line control. Some of the previously described control instruments can be used to regulate the thyratron tubes in this system.

Program Controls

When it is desired to control a furnace to a definite program, such as an annealing cycle, a series of heating-rate and

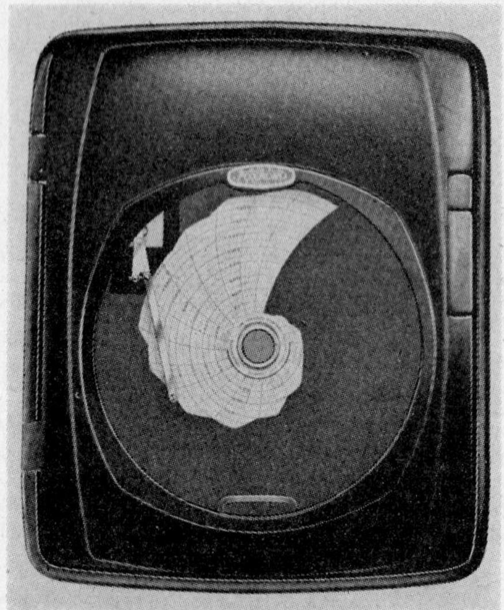

FIGURE 46.　*Control of Temperature Cycle, Utilizing a Cut Cam Arrangement (Courtesy, Foxboro Co.)*

holding-timer clocks can be connected into the control circuit. These can be manually preset for the desired cycle. As each portion of the cycle is completed, the control is switched automatically from one clock to the next. The heating and cooling rates are controlled by timers actuating a motor which drives the temperature-setting mechanism up or downscale at a set number of degrees per hour. The number of changes that can be obtained in one cycle will be governed by the number of timing devices installed.

Another method, shown in Figure 46, is quite simple and accurate. On the soft circular metal disk is marked a series of circular lines of increasing temperature from center to circumference. Curved time lines are inscribed radially on the disk so that one revolution is equivalent to any desired time from 4 hours to 7 days. The disk can be marked with the desired program by rotating it manually and tracing with the following arm. A pair of tin snips and a file are sufficient to cut the disk with the desired accuracy. This cut-out acts as a cam and is driven by a constant-speed motor. The following arm, which is connected by a system of levers to the setting index on the controller, rides on the cut edge of this cam. As the rotation of the cam proceeds, the arm follows the cut-out cycle and transmitts corresponding motion to the temperature-setting index. The furnace temperature will then follow the control index. If desired, a series of disks can be cut, having various heating rates which can be used interchangeably for either heating or cooling cycles by reversing them.

Chapter 3

OTHER METHODS OF TEMPERATURE MEASUREMENT AND CONTROL

Resistance Thermometers, Mercury and Bimetallic Systems

For temperature detection, besides the thermocouple, several other means are used, particularly for the low temperature range. While the platinum-platinum, 10 percent rhodium thermocouple is used as the standard calibration for high-temperature measurement, the pure platinum resistance thermometer is used for the low-temperature range, to approximately 1000°F. The operation of the resistance thermometer depends on the variation with temperature of the resistance of an electrical conductor. The material used for this type of thermometer must have an electrical resistance which varies with temperature, but a high coefficient is not necessary since accurate calibration is possible. The most important properties of the material to be used are reproducibility, constancy of resistance and a simple temperature-resistance relationship. Platinum is the metal most widely used, while nickel and copper are used for industrial purposes at lower temperatures. The fine resistance wire may be wound in a spiral form on a mica support. This delicate coil is then enclosed in a glazed porcelain or quartz tube. For industrial use, an outer metal protection tube is added. The various instruments for measuring the change in resistance of this unit include the Wheatstone bridge, the Thompson bridge, the potentiometer and the differential galvanometer.

The resistance thermometer is useful in laboratory work because it is capable of high precision, but it is somewhat fragile. Poor contacts or contaminated wire may cause great errors. Whereas the thermocouple only indicates the temperature at its hot junction, the resistance thermometer indicates the average temperature over its entire length. The thermometer bulb may be constructed to hold any desired length of resistance wire. Another advantage of this instrument is that no cold-junction correlations are necessary.

Figure 47 illustrates a simple and efficient resistance-thermometer control device. This open view shows the arrangement of an alternating-current Wheatstone-bridge control circuit. The temperature-setting dial is connected directly to the slide wire and is adjusted by a knob in front. As the temperature approaches the desired value, the voltage through the resistance on the bulb side of the bridge begins to balance the voltage through the previously set slide-wire resistance. When the temperature is reached and circuit balance is obtained, the relay kicks out and the red indicator light goes on. As the temperature drops, the relay makes contact and the green indicating light flashes on. In order to eliminate any changes in bulb calibration due to temperature variations or changes in lead length, three lead wires are used. One lead ties in directly with the power tube while one goes to the variable section of the bridge circuit. The third or compensating lead, which is connected to the same side of the bulb as the first lead, goes to the slide-wire side of the bridge. Two tubes are used in this circuit, one for amplification and for power in the bridge and another to supply power for the relay. This controller has only one moving part, the relay, and, therefore, requires practically no maintenance. It is used in controlling ovens, distillation processes, quenching oils and coolants. The resistance

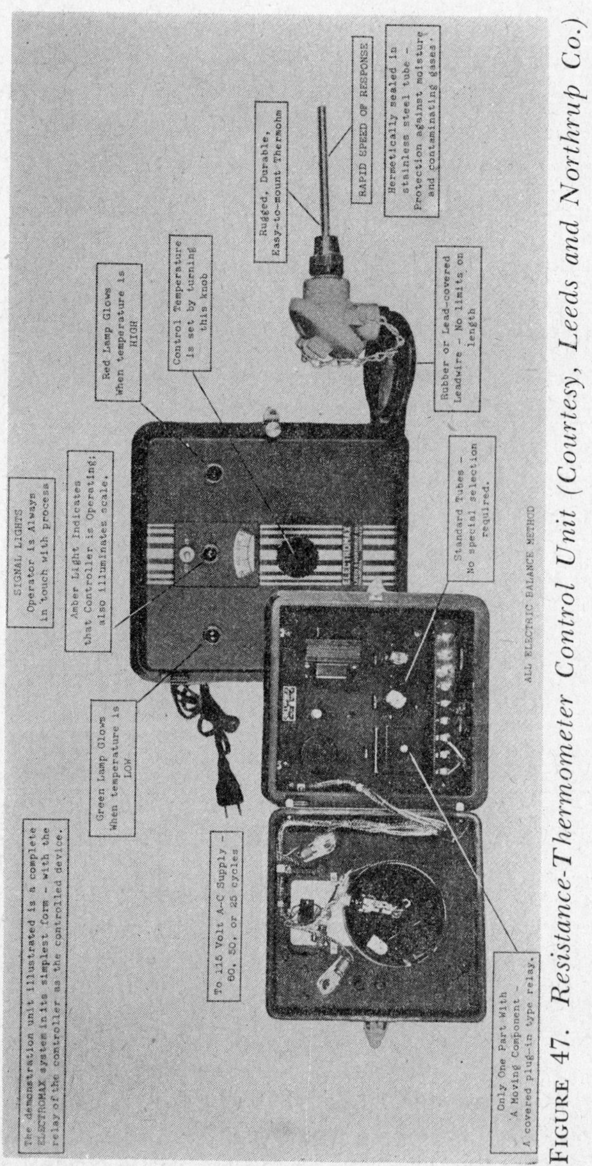

FIGURE 47. *Resistance-Thermometer Control Unit (Courtesy, Leeds and Northrup Co.)*

method of control can be utilized with the previously described recorders and controllers.

For close temperature control in the laboratory, the furnace element itself is utilized as the temperature-measuring element. The electrical resistance of the sealed platinum furnace element is calibrated against temperature and, by means of accurate relays, the temperature can be maintained to 0.5°F.

Another method of controlling the lower temperatures, up to 1000°F., is by means of a flat, spiral actuating tube which can be used with a mercury-filled thermal system. The mercury-filled bulb is immersed in the medium to be controlled, the changing temperatures causing the mercury in the bulb to expand or contract to actuate the controller. The expanding mercury causes the spiral measuring element to uncoil. This uncoiling action is indicated by a needle on

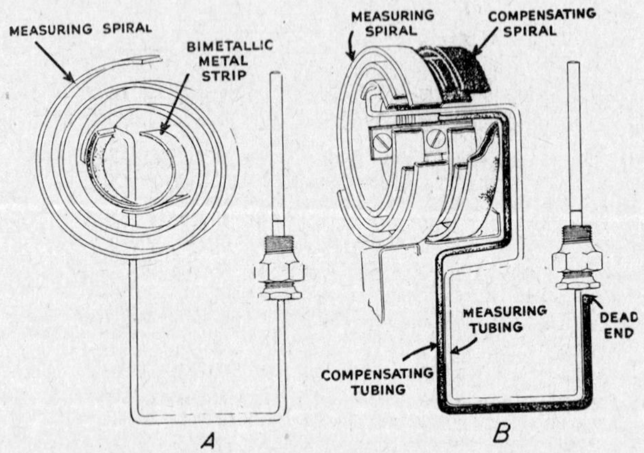

FIGURE 48. *Mercury-Filled Thermal System; Showing Method of Compensation*

a scale calibrated in degrees. However, any variations in temperature along the tube leading from the bulb to the controller will affect the operation of the controller. If the tube is relatively short and the temperature fairly uniform, then the only compensation required is for temperature variations of the instrument case. This temperature variation would affect the calibration of the actuating spiral. A method of compensation, illustrated in Figure 48 (A), consists of connecting the actuating spiral to a piece of bimetallic strip which will constantly correct for the instrument-case temperature. The action of the bimetallic spiral is opposite to that of the measuring spiral so that as the temperature of the instrument changes the net effect on the temperature-indicating needle is zero. The case and capillary compensation technique, as shown in Figure 48 (B), is required where relatively long tubes are necessary. The compensating spiral and tube are similar to the actuating element except the tubing dead ends at the measuring bulb. This unit is connected in the opposite direction to the operating spiral which, in turn, is attached to the pen arm. Thus, only the true temperature at the bulb is permitted to be recorded. In converting this unit into a controller, an electronic unit, similar to the Brown Electro-Vane, may be inserted. A schematic diagram of this unit is shown in Figure 49. This element is somewhat similar to others described in that an actuated metal vane passes between two oscillator coils, upsetting the state of oscillation and changing the amount of current flowing in the pickup circuit. This change of current actuates the power relay. The movable vane is connected directly to the spiral element and has a narrow zone of control action.

In addition to the resistance pyrometer and the mercury-filled thermal systems, good control can be secured in the low-temperature ranges by means of a nonindicating bimetallic-element controller. The bimetallic element is composed of two dissimilar metal strips fastened so that their surfaces are together. This unit is in a spiral form and as it is heated, the one metal will expand more than the other causing the spiral to open up. This element is enclosed in a protective tube which is placed into the medium to be controlled. The entire control mechanism can be placed at the point of temperature control since the protection tube acts as a support for the small unit. The bimetallic element transfers its motion to a mercury switch which is connected so as to operate the desired input controls. The temperature

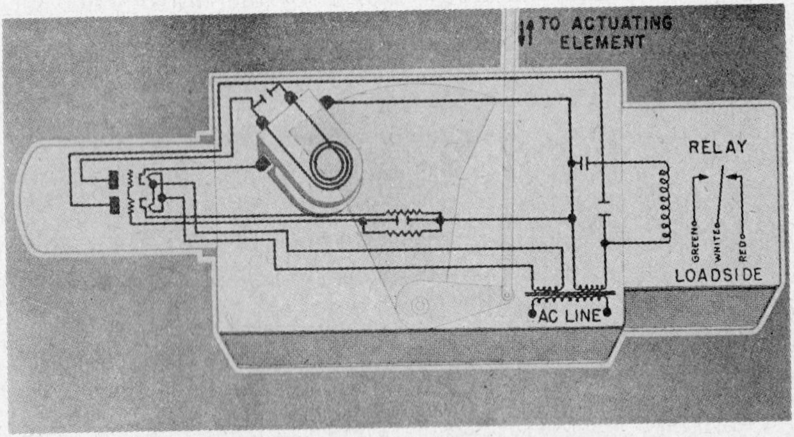

FIGURE 49. *Schematic Wiring Diagram of the Electr-o-Vane Unit (Courtesy, Brown Instrument Co.)*

may be set at the required value by moving a setting lever on the temperature scale. This changes the relative position of the mercury switch to the bimetallic element. A differential-adjustment lever can be regulated to give satisfactory temperature control.

Radiation Pyrometers

It is well known that as the temperature of a body increases more radiant energy is emitted. Utilizing this fact, the radiation pyrometer was developed. This type of pyrometer need not be heated as does the thermocouple hot junction, but is only sighted on the object to be measured. The radiation pyrometer covers a wide range of radiation and includes both heat and light. This radiation is converged on the hot junction of a small thermopile. The temperature rise of this junction is dependent on the rate at which the energy falls on it. The rate at which a heated object emits radiant energy depends on its temperature. This energy rate, governed by the Stefan-Boltzmann law, is proportional to the fourth power of the absolute temperature of the heated object. The temperature of the cold junction of the thermopile in the radiation pyrometer is not controlled. The hot and cold junctions are fairly close to each other and are equally affected by room-temperature variations. The radiation pyrometer is designed to shade the cold junction from the radiation received from the heated object. For greater accuracy, a thermopile is used in this instrument. This consists of a group of fine thermocouples connected in series, with their hot junctions adjacent to each other. The radiant energy affects the hot junction of this thermopile to raise its temperature, but to a much lower value than the actual temperature of the heated object.

Cold objects appear black to the eye because the light falling on them is obsorbed leaving little to be reflected to the eye for observation. Soot or lampblack, having practically no reflective power, are good examples. A similar effect is observed when looking through a small opening into a large, cold furnace muffle. The opening appears black because any light that gets through is absorbed by the inner muffle walls. If heated, the lampblack as well as the small muffle opening will emit radiant energy at a maximum rate. Such a body that emits radiant energy without hindrance is known as a "blackbody" or "perfect radiator."

In a large number of industrial heating processes, perfect-radiation conditions prevail; therefore, no temperature corrections have to be made for the radiation pyrometer. Muffle and box furnaces approximate perfect-radiation conditions so that any cold steel placed within will, after a given heating time, become indistinguishable from its surroundings. If a large clean piece of steel is placed in a hot furnace immediately it will appear quite hot due to the reflection of the hot furnace walls on the steel surface. Similarly, if a mirror were placed in the hot furnace, immediately it will appear just as hot as the furnace. Therefore, care must be exercised when checking the temperature of steel in a furnace to secure perfect radiation conditions. When the steel becomes coated with iron oxide, the reflection of the surface is greatly decreased. However, when the steel is removed from the furnace its apparent temperature is reduced because some surface reflectivity is still present.

Since optical and radiation pyrometers are calibrated under perfect-radiation conditions, a correction must be applied to the observed reading to compensate for reflection effects. Naturally a cleaned molten metal surface will require a greater correction than a piece of oxidized steel at the same temperature. If only temperature control of a process is required, the emissivity correction is not necessary since uniformity can be checked, knowing that the temperature will always read lower by the same amount.

The emittance or emissivity is defined as the ratio of radiant energy emitted in unit time by unit area of a body to that emitted by a perfect radiator at the same temperature. Some values for the emittance of various materials are shown in Table 3.

The emittance values are greater at higher temperatures for some materials. This occurs because the shorter-wavelength radiation at higher temperatures becomes more prominent, and since these radiations have different absorption characteristics, the emittance will change. The emittance values may also vary somewhat due to particle size, but if accurate values are desired for a given material a special check can be made. This is done in most cases by temporarily inserting a thermocouple into the sample specimen and checking its actual temperature. This temperature is then compared with the temperature reading from the radiation instrument.

To measure the accurate temperature of a material with a

TABLE 3

Various Metals and Their Emittance Values

Material	Temperature °C.	Emittance
Aluminum (oxidized)	200	0.11
	600	0.19
Brass (oxidized)	200	0.61
	600	0.59
Cast Iron (oxidized)	200	0.64
	600	0.78
Copper (oxidized)	200	0.6
	1000	0.6
Iron (oxidized)	100	0.74
	500	0.84
	1200	0.89
Lead (oxidized)	200	0.63
Monel (oxidized)	200	0.43
	600	0.43
Nickel (oxidized)	200	0.37
	1200	0.85
Steel (oxidized)	25	0.80
	200	0.79
	600	0.79
20% Nickel 25% Chromium 55% Iron	200	0.90
	500	0.97

radiation pyrometer, the observed millivoltage of the instrument is divided by the total emittance. This yields the corrected millivoltage which can be converted into temperature from the temperature-millivolt curve of the radiation instrument. For radiation pyrometers, calibrated to read in terms of temperature, the observed temperature must first be converted into millivoltage, using the temperature-millivoltage calibration curve of the instrument.

This radiation measurement gives a better average temperature reading than a thermocouple since it covers a larger area. By sighting close to the heated material, the area covered is approximately the same as that of the instrument tube itself. More coverage is secured by moving farther away from the temperature source.

In order to secure an accurate temperature measurement it is important that all of the radiant energy received by the radiation-pyrometer head comes from the heated object only.

FIGURE 50.   *Phantom View of the Brown Radiamatic Instrument (Courtesy, Brown Instrument Co.)*

strument. When the heated object is too small to fully cover the opening of the radiation receiver, the indicated temperature will be lower than the actual since the calibration is based on full coverage. However, small instruments of the radiation type are being developed to check temperatures of individual pieces during induction heating. A phantom view of the Brown Radiamatic unit is shown in Figure 50. The radiant energy from the heated object is Also, there must be sufficient area of the heated object to insure a complete coverage of the front opening of the in-

collected by the lens 2 and focused to strike the hot junction of the thermopile 3. This thermopile is composed of a group of fine thermocouples arranged in a circle with their hot junctions forming a rosette in the center. These couples are connected in series so that a small temperature increase at the hot junction produces a considerable millivoltage output. The cold junctions of the fine thermocouples make good thermal contact with the housing of the instrument, but changes in temperature of the unit will produce variations in cold-junction temperature, giving an inaccurate temperature measurement. To eliminate this difficulty, automatic compensation is secured, up to approximately 250°F., by the use of a heat-responsive nickel coil 1, connected in the thermopile circuit. In applications where temperature variations of the instrument may be quite considerable, air or water cooling is utilized. A sighting lens 4 is used to set the tube in the proper position to receive rays from the specific area. These units are factory-calibrated under perfect-radition conditions, but when used to check temperatures during forging and rolling of plates or rods, the instrument must be recalibrated to suit these individual conditions. This adjustment can be made by moving the pinion and diaphram indicated by 5. In this instrument two types of lenses are used, depending on the temperature range desired. Pyrex heat-resistant glass is used for temperatures of 2300°F. and above since it permits a ready passage of the shorter wave lengths predominating at these high temperatures. For temperatures of 800 to 2300°F., the fused quartz lens allows fice in the tube to reduce the receiving area, the temperature the longer wave lengths to pass. By inserting a smaller orifice scale may be shifted from low to high within the total range of the instrument.

Another type of instrument (fixed focus) has a mirror behind the thermopile. The energy striking the curved mirror is focused on the hot junction. A cover glass on the front end protects the mirror from dirt and dust.

A low-temperature-range, 100 to 600°F., radiation unit, designed by Brown Instrument Co., utilizes a calcium fluoride lens and a concave reflecting mirror. The fluoride lens, made from artificially grown crystals, has the ability to transmit the longer wave lengths of radiant energy as emitted by a low-temperature body. To maintain a constant cold-junction temperature, a special resistance-thermometer controller and heater coils are incorporated. The temperature of the resistance-thermometer winding (close to the cold junction) regulates the power on the heater coils (located around the jacketed radiation tube). Thus, the unit is responsive only to the radiation focused on the thermopile. An arbitrary calibration of 0 to 100 permits the use of the instrument under a variety of emissivity conditions. In most cases, only the consistent reproduction of known temperature conditions is required.

The radiation pyrometer is a very quick-acting instrument. It requires approximately 2 seconds, and is used in measuring the temperature of hot objects either moving or still. This rapid response is possible since the radiant energy is focused on the very fine hot junction of the thermopile which, having low thermal capacity, heats quite rapidly. Since no actual contact is necessary, the entire tube assembly is sealed so that dust, dirt and fumes cannot affect the inner components and long life is assured. But care must be exercised in setting this instrument to obtain accurate results. Excessive flame, smoke, dust and water vapor will yield inaccurate temperature readings due to the absorption of some of the longer wave lengths of radiant energy. Recalibration must be resorted to if measurements are to be made under poor radiation conditions. Rapid checking with an optical pyrometer is possible when the radiation pyrometer is used under conditions of a perfect radiator.

Two methods of installation are widely used for securing best results with the radiation-type instruments: closed-end and open-end sighting tubes. In cases where there are no perfect radiation conditions, the radiation pyrometer can be fitted to the opening of a closed-end ceramic tube. This tube is so long that the closed end can be inserted into the hot zone of the furnace. This portion of the tube heats up to the furnace temperature and the instrument needs to measure only the radiation emitted from the bottom and sides of the tube (a good radiator). However, one possible source of error is the collection of smoke or other absorbing

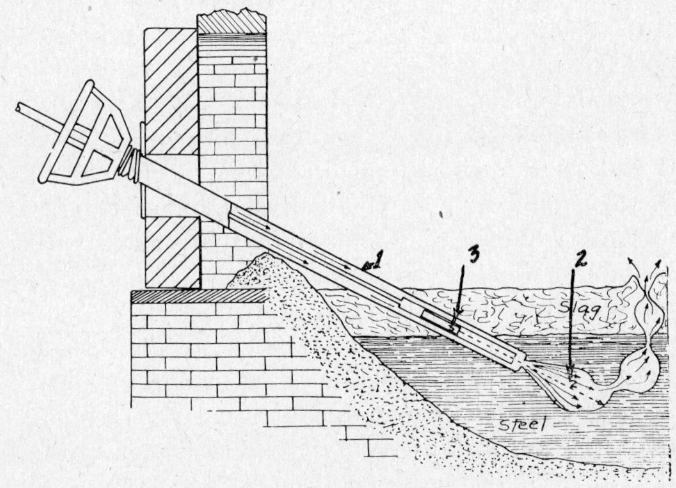

FIGURE 51.   *Leeds and Northrup Blowing-Tube Pyrometer. Rayotube 3 Inside of the Pyrometer Tube 1 Receives Radiant Energy from the Liquid Surface of Air Bubble 2 (Courtesy, Leeds and Northrup Co.)*

gases within the tube. A Fyrstan ceramic tube, being quite dense and impervious to gas, works quite satisfactorily.

The open-end tube is used for checking the actual temperatures of the material in the furnace. For measuring or controlling the temperature of steel billets or slabs in a reheating furnace, the tube is inserted in the top of the furnace to almost contact the surface of the steel. A wide flange or collar at this hot end prevents reflections within the furnace from affecting the radiation unit. To prevent the tube from becoming filled with smoke, clean gas or air enters at the top and is gently forced down the tube. Another application is to sight the open-end tube at a point on the roof of an open-hearth furnace. The radiation instrument can be used to control the open-hearth temperature for maximum roof life.

For measuring actual molten-steel temperatures in the open-hearth furnace, a radiation-type instrument, as illustrated in Figure 51, is built by Leeds and Northrup. The radiation receiver is fitted inside a low-carbon tube at approximately 13 inches from the open end. The full tube length is 7 feet, with the electrical and air connections at the cold end. The end with the radiation receiver is pushed through a small hole in the wall and immersed into the liquid steel. The air flow is maintained at sufficient level to keep the radiation receiver cool and to force back the steel from the immediate opening of the tube. Thus, it is the radiation from the air pocket at the end of the tube that is measured. Due to the oxidation of the metal by the air, this tube gives a higher temperature reading than a platinum couple. The use of nitrogen in the tube yields a temperature reading about the same as that given by the platinum couple. The high reading obtained when using an air stream can be adjusted by compensation with the checking unit. With the radiation receiver located close to the molten steel, difficulties due to any warpage of the long sighting tube and change in size of the opening are reduced. This proximity to the molten metal also improves the response of the instrument. The radiation-limiting orifice is located on the immediate front of the radiation receiver so that its change in dimension during operation is not too great. Thus, if the submerged end of the tube does partially freeze over, a temperature check can still be made. Care should be exercised not to leave the unit in the furnace too long as overheating and excessive corrosion will result. Therefore, rapid recording on the electronic Speedomax (2-second full-scale travel) is used. During a normal immersion, the temperature rise of the radiation receiver is approximately 175°F. For rapid checking and calibration, an additional test unit is clamped to the side of the immersion unit and both are sighted at some fairly constant-temperature portion of the furnace such as the back wall. The outputs of both units are connected to

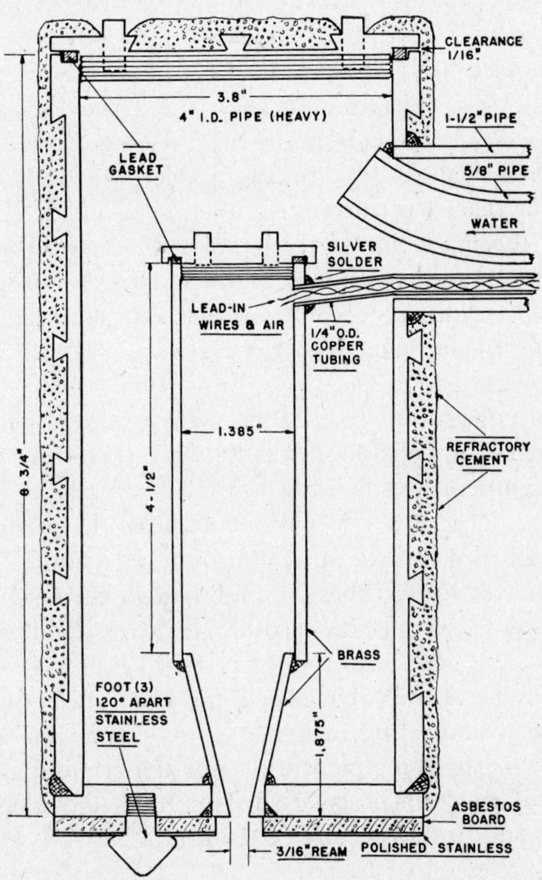

FIGURE 52. *Construction of a Special Radiation Pyrometer for Measuring Steel-Slab Temperatures (Courtesy, United States Steel Research Laboratory)*

a galvanometer which balances when both instruments are the same. A potentiometer dial on the control panel of the immersion unit can be adjusted to correct any disagreement.

In using a radiation instrument to measure the temperature of moving ingots, slabs and sheets, a short open tube is used with either air or water cooling. The instrument is suspended vertically over the mill rolls and as the heated steel passes, its temperature is quickly recorded. The water cooling of this instrument tends to decrease its speed. In order to accurately determine the surface temperature of steel slabs in a reheating furnace, the United States Steel Company Research Laboratory has devised a holder as illustrated in Figure 52. In this holder, the radiation pyrometer is placed in the inner tube and water is circulated between the inner and outer jackets. The assembly is held by a 10-foot handle traversed by the cooling water and lead wires. The stainless steel legs are placed on the slab surface within the furnace and readings are obtained within 3 to 5 seconds when a high-speed recording potentiometer is used. This rapid measurement eliminates any cooling of the slabs. Such construction enables this radiation receiver to withstand furnace temperatures of 2400°F. for 30 seconds.

Another successful application of radiation pyrometry is for temperature control of a continuous strip-annealing furnace. The radiation tubes are mounted in the roof and are sighted into closed ceramic tubes which extend down to practically the middle of the furnace. The closed-end sighting tube reduces the vapor effects and also reflectivity differences due to annealing in bright or oxidizing atmospheres. Although the strip temperature is not actually obtained, the hot end of the tube is positioned in the same zone as the strip, assuring uniform heat control.

Photoelectric Instruments

Two types of photoelectric cells are being used as temperature detecting devices; the cesium vacuum cell and the self-generative or barrier-layer cell. The first type, sensitive to the red end of the visible spectrum and a portion of the infrared, will emit electrons, depending on the relative amount of radiation striking it. With this vacuum cell connected in the proper circuit, the emitted electrons will affect a current flow proportionally to the amount of recipient radiation. The second type of cell is the same as those used in photographic meters, with special precautions to secure good response to low radiation intensities. No additional electrical power is necessary to operate this barrier-layer cell, and it has a radiation-sensitivity curve comparable to that of the eye.

Utilizing a blowing-tube arrangement similar to that designed for the radiation instrument, another unit has been developed which measures the temperature of molten steel by means of a calibrated barrier-layer cell. This cell, however, instead of being placed close to the hot tip of the tube, is fastened at the air-inlet end. Light-control diaphrams spaced evenly along the inside of the tube as well as the immersion-end opening permit only a true beam of radiation to strike the cell. Since the current output of this instrument is very small, special shielded coaxial cables are necessary. For these blowing-tube instruments clean, dry air must be available to secure the proper results as any dirt collecting on the cover glass of the temperature detector will decrease the indicated temperature. The front protective glass on the barrier-layer cell prohibits the passage of infrared rays and thus prevents the overheating of the sensitive cell surface. If very-high-speed records are desired, the instrument may be attached to a recording milliammeter with an operation speed of less than 1 second. For normal work, one of the modern electronic potentiometer recorders, with full-scale travel in 4 seconds, is satisfactory. For initial calibration, the temperature versus cell current-output relationship must be obtained. Subsequent checks require only the comparison of the cell output with that of a platinum thermocouple at a uniform temperature of approximately 2700°F.

A system has been developed by the Brown Instrument Co., called the Optimatic, which utilizes vacuum-type photoelectric cells to measure automatically the intensity of light emitted from a heated object. The schematic outline of this system is shown in Figure 53. This unit is sighted on the hot object and the radiation, principally dark-red and short-

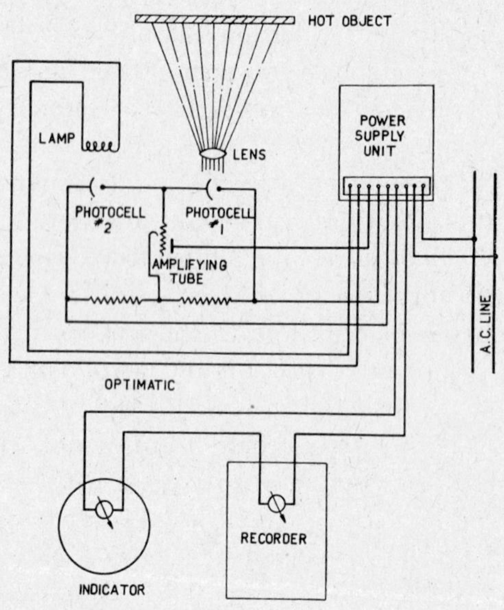

FIGURE 53. *Schematic Outline of the Brown Optimatic System (Courtesy, Brown Instrument Co.)*

wave infrared, is converged through a lens to the first photocell. As illustrated, this system utilizes a bridge circuit with an amplifying tube as a detector and balancing device in place of a galvanometer. As the radiation strikes the first photocell (1) and changes its electrical resistance, the plate circuit of the amplifying tube indicates an unbalanced condition. This unbalance is immediately reflected back to the power supply which increases the current to the balancing carbon-filament lamp. The brightness of this lamp grows sufficiently intense to affect the second photocell (2) so as to balance the circuit. The current supplied to the lamp passes through a milliammeter recorder in the instrument which is calibrated in terms of temperature. With this system, temperature records can be secured in approximately ½ second. The use of two photocells improves the stability and reduces the harmful effects of voltage fluctuations. However, both cells should be matched if best results are to be obtained. This method of temperature measurement is easily adaptable to steel hot-rolling operations where the object moves quite rapidly, and the measurements can be recorded for future reference. Figure 54 is a phantom view of the sighting tube and of the photocell arrangement.

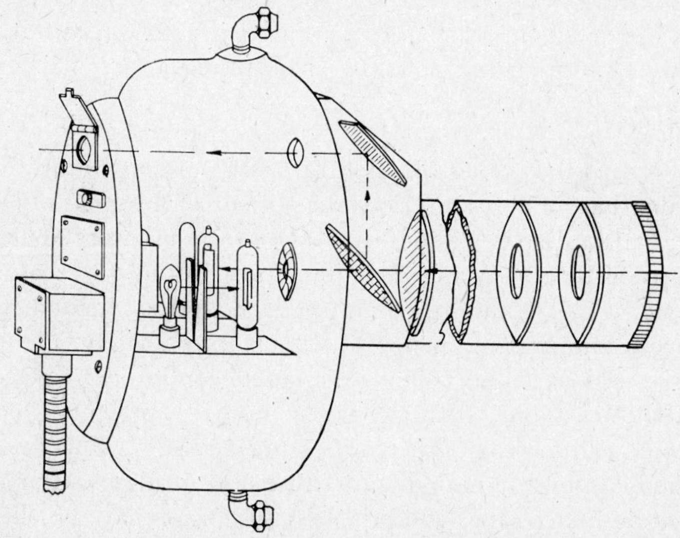

FIGURE 54.   *Phantom View of Optimatic (Courtesy, Brown Instrument Co.)*

For high-speed continuous-line recording of temperatures, as detected by the barrier-layer or vacuum photocell-type instrument, a special milliammeter, utilizing an internal photocell circuit, can be used. This type of instrument has a full deflection speed of approximately ½ second. A small, free-swinging detecting galvanometer, with attached mirror, is connected directly to the temperature-detecting element. A constant light source is focused on this galvanometer mirror and reflected back through a dividing prism to affect two photocells controlling a triode vacuum-tube circuit. Thus, any movement of this small galvanometer is amplified through the photoelectric circuit to operate a large, electrically damped, recording galvanometer. The needle of this large galvanometer consists of a hollow capillary tube which transports the ink from a large, open well to the curved pen tip. This tip is curved downward to a vertical position and rests on the paper chart very lightly. With this arrangement, the friction between the pen and paper chart does not influence the accuracy of the instrument as ample power is supplied to move the recording pen.

Optical Pyrometers

Perhaps the earliest method of checking temperature was estimating by eye the relative brightness of a heated object. There is a direct relationship between the intensity of light given off by a body and its temperature, but as the measurement based on this relationship is an optical method, the lowest temperature measurable is approximately 1200°F. Since radiant energy is being measured, conditions of perfect radiation, as previously described, must be present. Dense flame will increase the temperature reading while smoke clouds absorb radiation and cause low readings. Carbon dioxide, water vapor and other invisible gases have no effect. When necessary to sight an optical pyrometer into a furnace through a window, the proper temperature correction must be made. The loss in light intensity in passing through a glass window is approximately 3.5°C. at 600°C. and 10°C. at 1200°C. When properly used, the optical measurement is quite accurate and is about the only means of determining temperatures of 1500°C. and above. The big disadvantage of this type of instrument is that it is not recording and must be operated manually.

The small, nonrecording, portable optical pyrometers have been used quite extensively as a means of rapidly determining pouring, casting or rolling temperatures in the steel mill and foundry. Two types are popular, the disappearing filament and the optical wedge. The first matches the total brightness of the heated object with that of an adjustable, temperature-calibrated lamp, while the second utilizes a lamp of standard intensity, the matching being accomplished by means of a graduated optical wedge inserted in the path of the radiation from the heated body.

The disappearing-filament (Leeds and Northrup) optical pyrometer consists of two electrically connected sections, the telescopic sighting unit and the calibrated box for measuring the potential drop through the lamp filament. In operation, the telescope is sighted and focused on the hot object. The presence of a red filter inserted in the optical system gives monochromatic radiation to the eye and permits a more accurate measurement. By inserting a darker-red filter, the instrument is shifted to the high-temperature range. Also, within this optical system is a small tungsten lamp, which has a single-strand filament with a sag in the middle. This section of the filament is superimposed on the image of the heated object, and the knob on the control box is turned to pass more current through the filament. The current is adjusted until the brightness of the filament matches and becomes indistinguishable from the heated object. If the filament is at a lower temperature, it appears darker than the object while if hotter, it appears brighter than the background. The calibrated box supplies the current to the filament (dry cells and rheostat are used) and also measures the current flow through the filament. The older models utilized a milliammeter to measure the required current. However, the modern instrument utilizes the more accurate potentiometer measurement of the voltage drop across the filament. The scale, which is adjusted by the same knob that regulates the filament current, is calibrated directly in temperature degrees. The lamp in this instrument has its brightness calibrated against the current. After the current is adjusted to secure the proper filament brightness, the knob is pushed in to engage the potentiometer circuit. After rotating the knob till the galvanometer becomes balanced, the temperature is read on the scale.

The instrument utilizing the calibrated circular glass wedge (Pyrometer Instrument Co.) is a small self-contained unit. The sighting of the telescope and the method of changing the temperature range are the same as previously described. The small filament lamp in this instrument is held at constant intensity and is set at the calibrated point by moving a small rheostat knob until the required current registers on the ammeter scale. A small dry cell supplies the current. The filament of this lamp is superimposed on the field by means of a glass prism. The red filter yields a monochromatic image to eliminate any effects of color differences as only the relative brightness is measured. The lamp is brought into the line of sight as a small dot in the center of the field. This unit is sighted on the object, the button is depressed to light the lamp with one hand and the other hand rotates the graduated circular wedge. This wedge is moved until the background (heated object) equals the intensity of the bright central lamp spot. At this point of balance, the background and central dot become indistinguishable. If the pyrometer is set higher than the object temperature, the central spot will be brighter than the background and when lower, the central spot will be darker. A temperature scale is attached to the calibrated circular wedge.

To measure the temperature of very small objects, such as incandescent lamp filaments, and for laboratory work, the Pyrometer Instrument Co. has developed a micro-optical pyrometer. The optical system magnifies the heated object approximately twenty times at a distance of 6 inches. This instrument is similar to a microscope and, therefore, must be held stationary and free from vibration. Exact focusing of the heated object and calibrated lamp is accomplished by means of worm gears. The temperature measurement is made accurately by adjusting the amount of current to the calibrated lamp with coarse and fine resistance knobs. Once the calibrated lamp matches the brightness of the image of the heated object, the amount of current is read on an ammeter. Calibration curves are furnished with each standard lamp. Four flashlight cells are used to supply the required current.

These optical instruments are calibrated by sighting on a standardized ribbon-filament lamp and comparing the resultant indicated temperatures. Under proper conditions of radiation a careful observer can obtain an accuracy of plus or minus 3°F., and a group of trained observers should be within 10°F.

These instruments also require perfect radiation conditions for accurate temperature measurement. The inside of a furnace may approach this condition, but variations may be considerable in measuring highly reflecting surfaces in the open. To eliminate this variable in radiation conditions, the Pyrometer Instrument Co. has developed an optical pyrometer which measures the characteristic temperature color emitted from the heated object. The color spectrum emitted from the heated material depends only on temperature and is not affected by changes in surface reflectivity. Figure 55 shows the operating details of this instrument. The bichromatic, red-and-green color wedge C is adjusted in the path of the radiation from the object to the eye. Only varying ratios of red and green light are transmitted as this wedge is changed in thickness. When red and green are mixed in a definite ratio, the result appears white to the eye. The color wedge is moved to balance the red from the heated object and produce a white background in the scope.

This white background is matched to that produced by the standard lamp. The lamp is reflected into the scope vision by means of a partially silvered and cemented glass cube E. After the color adjustment is made, wedge D (being color

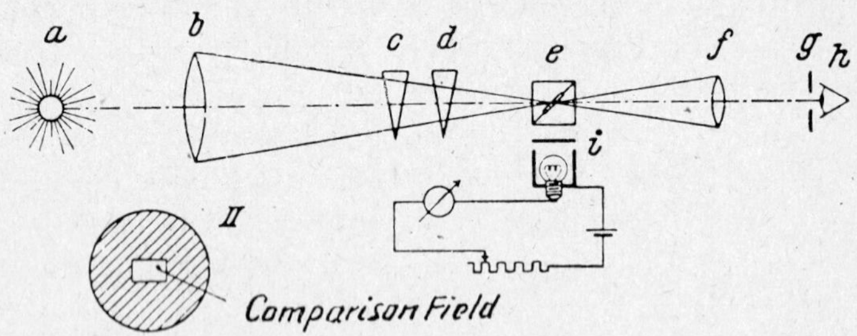

FIGURE 55.   *Diagram of a Portable Color Optical Pyrometer
(Courtesy, Pyrometer Instrument Co.)*

neutral but changing density) is moved to match the intensity of light from the heated object with that of the standard lamp. The scale adjusted by wedge C shows the true temperature of the heated object. The second scale, adjusted by wedge D, gives the normal optical temperature, and is affected by the surrounding radiation conditions. The difference between these two temperatures is an indication of the radiation conditions existing on the surface of the heated object.

One of the simplest color optical pyrometers is illustrated

FIGURE 56.   *Slide-Rule-Type Color Optical Pyrometer*

in Figure 56. This instrument has the appearance of a slide rule. The light filter is a composite wedge transmitting an increasing ratio of red to green color from one end of the instrument to the other. If the light passing through the filter is predominantly red, the object will appear red and if green is in excess, the heated body will look green. When the two colors pass through the filter in an equal degree, the body will appear white or practically colorless. While sighting the body through the movable slide and filter, adjustment is made by moving the slide to the position where both colors fuse and the object becomes white in appearance. The temperature is read on the calibrated scale. These color-detecting instruments would be excellent if they were equipped with some mechanical means of detecting the proper balance, since all observers' eyes do not have the same color sensitivity.

In the pouring of liquid steel and cast iron, bright regions can be observed in the stream which gives noticeable higher temperature readings for optical pyrometers. These varia-

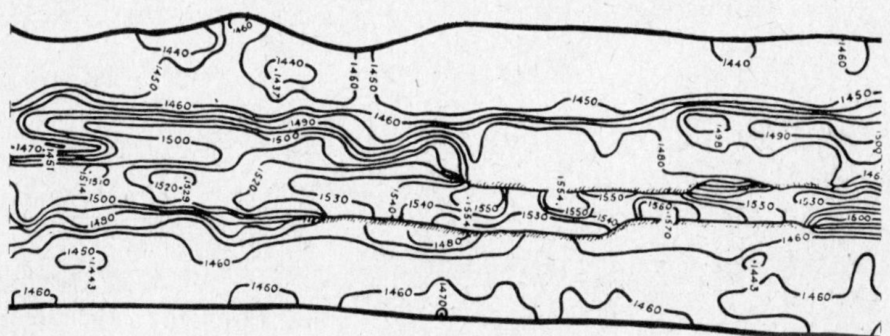

FIGURE 57. *Temperatures in Centigrade Degrees of a Molten Stream of Steel as Determined by Photographic Means*

tions are due to the presence of slag (having a higher emissivity than the metal), and also to the irregularities and folds on the metal surface. It is also possible that the formation of a light oxide film on the clean metal surface gives rise to bright spots. In the case of folded metal surfaces, these depressed areas will approach perfect radiation conditions and will appear brighter. To observe the brightness variations, an investigation was conducted by J. A. Hall in England, using photographic means. A movie camera was sighted on the stream and, with proper filters, pictures were taken on one half of the film. On the other half, a calibration was made, using the standard tungsten strip-filament lamp set at various temperatures. With the aid of the calibration curve of the film, the film densities were measured with a microphotometer and converted into temperature readings. Figure 57 illustrates the temperature variations in a stream of molten iron during pouring. This method of measurement is not too accurate and requires considerable standardization, however, it shows that the photocell pyrometer covering a large area will read approximately 5° to 10°C. higher than the regular disappearing-filament optical pyrometer.

Chapter 4

APPLICATIONS, CARE AND MAINTENANCE
OF PYROMETERS

In the steel mill, most of the pyrometer applications are for close control of various types of operations ranging from the blast furnace to the small tool heat-treating furnace. For these installations, it has been found advisable to use the recording-type controller, since the charts can be studied and rechecked to improve the operation of the furnace and thus improve the quality of the products. When an instrument is installed, care should be exercised in positioning the temperature-detecting unit so that the best performance can be realized. This requires a knowledge of the fundamentals of heat transfer, combustion, electrical measurement and furnace design. Wherever possible, the proportioning type of control should be utilized, since the on-off type does not yield sufficient control of the temperature variable in manufacture. Another point of importance is that for accurate control, any pyrometer unit should be checked at least once a month.

Thermocouples

These are most widely used for temperature detection, being applicable over a temperature range of $-300°$ to $2800°F$. Reheating, patenting, normalizing, spheroidizing and annealing furnaces use this method of temperature detection. The higher-temperature furnaces require a platinum couple which has a fairly high initial cost, but the couples for lower temperatures are relatively inexpensive. The thermocouple has a relatively short life, but it can be replaced easily. To insure its accuracy, it should be fre-

TABLE 4

Composition and Properties of Thermocouples

Material	*Composition*	*Polarity*	*Identification*
Iron	Pure	Plus	Highly Magnetic
Constantan	55% Copper	Minus	Nonmagnetic
	45% Nickel		
Chromel	90% Nickel	Plus	Nonmagnetic
	10% Chromium		
Alumel	95% Nickel	Minus	Slightly Magnetic
	5% Aluminum		
	Manganese		
	Silicon		
Platinum-Rhodium	90% Platinum	Plus	Slightly Harder
	10% Rhodium		than Platinum
	or		
	87% Platinum		
	13% Rhodium		
Platinum	Pure	Minus	Very Soft

quently checked both visually and by comparison tests. The hot junction of the test couple is placed next to that of the control couple and read by means of a portable potentiometer. This will detect any contaminated or badly oxidized couples. Couples in use that become brittle are usually out of calibration. Platinum couples may be quite brittle, but appear perfect to the eye. It is possible to reclaim these couples by cutting back the brittle end and rewelding, however, the reconditioned couple should be recalibrated. Considerable care must be taken in protecting couples from contamination. Platinum-platinum rhodium and Chromel-Alumel couples cannot be used in reducing atmospheres. When these couples must be used under such conditions, a good protection tube is required to keep out the harmful gases and, in the case of platinum, double tubes must be used. Even a small amount of grease or dirt may be sufficient to contaminate this couple.

TABLE 5

Lead-Wire Color Codes

Company	Iron-Constantan		Chromel-Alumel		Platinum Rhodium-Platinum Compensating	
	Plus	Minus	Plus	Minus	Plus	Minus
Leeds and Northrup	White	Red	Yellow	Red	Black	Red
Brown Instrument Co.	White	Red	Yellow	Red	Black	Red
Bristol	Red	Green	Black	Green	Red	Black
Foxboro	Black	White	Black	Green	Black	Red

It is important to connect the couples properly at the lead-wire junction and to use the right compensating lead wire. Any temperature variations between the lead wire and thermocouple, in the case of mixed wires, would give erratic results. In fact, care must be exercised at all regions of temperature gradient to be sure that no countercurrents are produced. Table 4 shows the polarity and chemical composition of the wires of three important couples.

In Table 5 are some of the color codes for solid (Iron-

Constantan and Chromel-Alumel couples) and stranded (platinum couples) duplex lead wires.

The couples should always be inserted in the furnace in such a manner that the heat conduction along the wires is at a minimum. Since the hot junction is the measuring point of a couple, any excessive removal of heat from this point would indicate a lowering of its temperature. Also, the hot junction should be located at the high-temperature point in the furnace so that some degree of protection from overheating may be acquired. However, various positions of the couple should be tried and checked to determine the best location.

One particular disadvantage of the thermocouple is that it is restricted to stationary objects. As used in continuous furnace operation, only the temperature of the furnace is controlled. The speed of the material through the furnace then determines the temperature attained. However, the actual temperature of the heated material can be obtained using another method of measurment, radiation or optical.

Loose connections are a source of trouble which will lead to erratic control. For process annealing of coiled strip in the mill and also in the laboratory, polarity plugs are used to advantage. These connections are easily made and are secure. The plug is designed so that reverse connections cannot be made.

For industrial standardization of couples, the secondary method is recommended. The standard platinum couples can be obtained from the Bureau of Standards. At least two should be maintained and frequently rechecked by the Bureau. For checking an unknown couple, a small, one-inch tube furnace with satisfactory temperature range is sufficient. The hot junction is placed into a small hole drilled into a one-inch-round, pure-nickel bar. This two-inch-long bar has a similar hole in the opposite side to take the hot junction of the platinum couple. The bar, located in the center of the furnace, is used to maintain both couples at the same temperature. A series of hot-junction temperatures can be obtained, and the output of the unknown compared with that of the standard couple. A potentiometer is used to measure the output of both instruments. It must also be remembered to adjust for the cold-junction temperatures.

Radiation Unit

Since this temperature detecting unit does not come in contact with high temperatures, but is sighted on the hot object, its life is relatively long. Air or water cooling keeps the instrument temperature uniform and relatively cool. Any slight temperature fluctuations of the cold junctions of the thermopile are compensated by means of a temperature-sensitive spool of resistance wire.

Radiation pyrometers can be used for practically any application in which a thermocouple is used and, in some cases, is more flexible and adaptable than thermocouples. Since no contact is necessary and the reaction speed is high, moving-object temperatures can be measured quite easily. Good temperature control of the actual steel itself can be secured when this instrument is used on a continuous annealing furnace or a billet-reheating furnace.

The initial cost is quite high, but the pyrometer requires little maintenance. One important item is to keep the front receiving lens clean and free from dust and foreign particles. Any decrease in the effective aperture of the radiation unit will produce low temperature readings. In fact, anything that absorbs radiation between the temperature source and the instrument will give low readings. Fumes, steam or cool layers of carbon dioxide will absorb some of the radiation and, therefore, these rays will not be concentrated on the hot junction of the thermopile. By the use of proper lenses, however, the affected wave lengths of radiation can be absorbed almost completely by the lens material to reduce the effect of these gases. The radiation instrument must be temperature calibrated for the radiation transmitted by the lens. The Stefan-Boltzmann law does not apply to the selected radiation employed by the lens-type instrument nor to the mirror type using a protective window. For particularly low temperatures, about 300°F., mica or calcium fluoride windows or lenses may be used, since their absorption of radiation at this temperature is small.

For perfect radiation conditions, these instruments are checked and calibrated by the manufacturer, but various emissivity conditions necessitate extensive corrections in some cases. The interior of furnaces approach perfect radiation conditions and no corrections are required. For hot steel, with a covering of iron oxide, only a slight correction is required, but for bright surfaces, such as liquid steel or smooth rolls, the correction figure becomes quite large. Since, in most cases, it is difficult to secure accurate data on total emissivity, the radiation instrument can be used for control after the conditions have been previously established. Above 1500°F., this unit can be checked by means of an optical pyrometer. At lower temperatures, a thermocouple may be used for checking.

The most serious error to which radiation pyrometers are subject is that arising from stray radiation. Because of such a limitation, it is advisable to use the same object and sighting distance (or an equivalent ratio) in calibration as used in the actual installation. Primary calibration of this instrument consists of sighting the tube at a uniform-temperature body under perfect radiation conditions. The perfect radiator consists of a graphic closed-end tube having several diaphrams placed within to eliminate external-reflection errors. This tube is placed in a furnace having a uniform temperature distribution. The temperature of this graphite tube is accurately measured by a standardized platinum couple. A more rapid method of standardization utilizes a radiation pyrometer which has been previously calibrated as just described. The unknown tube is sighted on a uniformly heated source, and an immediate check is made with the standard instrument.

Since the radiation pyrometer covers a larger area than either the optical pyrometer or the thermocouple, it is more difficult to calibrate. Other errors are present which depend, to a large extent, on the type of the installation and the nature of the equipment.

Photoelectric Cell

This type of temperature-measuring unit has a relatively low installation cost, and very little maintenance is necessary. Here again the most important feature is to keep the optical system clean and free from foreign material. An optical lens and diaphram system is required to properly focus the light rays from the heated object on the surface of the cell. This instrument is also sighted on the heated body to obtain measurement; therefore, it is easily used for moving objects. Its reaction time is shorter than that of the previously discussed radiation unit, in which the radiation from a heated object increases about four times as fast as the absolute temperature. The radiation utilized in photo-electric instruments increases approximately 18 times as fast as the absolute temperature. Readings can be secured with this type of instrument in less than one second.

The output of this instrument is not appreciably affected by changes in the actual temperature of the receiver, but excessive heat, above 120°F., and moisture will cause deterioration of the cell. For continuous use, the cell must always be enclosed in a water jacket. A filter of heat-absorbing glass is usually inserted in front of the cell. This instrument is not used to any extent below 1800°F. and the current output is so low that an amplifier is needed.

The barrier-layer cell is used primarily for temperature measurements of the open-hearth roof and bath as well as for reheating and hot rolling steel products. Its calibration is quite similar to that of the total-radiation instruments.

Optical Pyrometers

These units, based on the photometer principle, are portable and easy to read. If the radiation conditions are good, such as inside a furnace, accurate temperatures may be secured with no correction needed for emissivity. Optical pyrometers measure the intensity of visible radiation and can be used on moving or stationary objects. Once the operator has obtained the knack of using this equipment, rapid readings can be obtained and the results are sufficient, in a number of cases, to eliminate the necessity of a permanent installation.

Optical pyrometers are used in the steel industry in the range of 1500° to 3000°F. This range covers such applications as patenting, hot rolling of rods, sheet and plate, and melting and pouring temperatures of an open hearth. If the heated object is small, it is more difficult to obtain the proper match in intensity particularly if the object is moving.

Here again smoke, vapors, dust, fumes and emissivity may affect the intensity of the visible radiation from the heated material so that care must be observed when such conditions are present. Such difficulties may be minimized by sighting down an open- or closed-end tube, whose end is located in the hot zone.

The battery used to supply the lamp current must be renewed quite frequently, but the most critical point is the lamp itself. When in constant use, the calibration of this instrument should be checked at least once a month for deterioration of the lamp filament. Routine mill checking is accomplished by sighting the optical pyrometer on a flat tungsten ribbon filament lamp which has been previously calibrated at the Bureau of Standards. The lamp should be completely enclosed to eliminate extraneous radiation. A precision potentiometer is used to regulate the amount of current going to the calibrated lamp. Precision resistors and storage batteries maintain the lamp at a definite brightness, its actual temperature being secured from the temperature-current calibration curve.

Millivoltmeters

These units are less expensive than the potentiometer-type equipment and are quite simple to operate. Constant improvements of the instrument regarding pivots, coils, high internal resistance and compensation have resulted in good trouble-free operation. The high internal resistance reduces the effect of the lead-wire length while the compensation eliminates the effect of temperature changes at the instrument.

As far as maintenance is concerned, dust, corrosion and vibration are the most important items to check. Dust in the pivots or between the movable coil and the permanent magnet will give erratic results in temperature measurement. Corrosion will give the same trouble while severe vibration may completely ruin the galvanometer movement by flattening or bending the pivots. The newer electronic models have been designed to give satisfactory performance in recording and controlling and their use is not restricted to any particular field. A portable potentiometer can be used to check the accuracy of this instrument, the connections being made at the regular lead wire terminals. By setting the portable unit at a given number of millivolts an equivalent reading should be obtained from the millivolt-meter-instrument scale.

For all measuring and recording instruments, proper care and lubrication are a necessity if satisfactory results are desired. While the instrument is in operation, inspection, cleaning, lubrication and repair should be done at least once a day by men specially trained in this type of work. Cleaning such delicate instruments is not just a matter of blowing out dust and dirt with compressed air. It means a careful wiping of each part. This is particularly true with electrical contacts and slide wires. A thin film of petrolatum will supply the necessary lubrication and will not interfere with electrical contact.

Potentiometers

These temperature measuring units have been widely accepted as the standard for accuracy and ruggedness. They are nevertheless susceptible to vibration which will affect the galvanometer action. The length of lead wire does not affect the calibration, but the action of the potentiometer may become somewhat sluggish if the length becomes excessive. These units can be used to measure the output of any voltage-type receiver: thermocouples, radiation or resistance units. The portable unit is a necessity for checking both thermocouples and instruments. It can be calibrated in either degrees or millivolts, but for checking purposes, where various couples are used, the millivolt calibration is preferred. The tables at the back of the book are used when the cold junction is kept at a constant temperature in an icewater mixture (32°F. or 0°C.). When a manually compensated instrument is used and the cold junction of the thermocouple is at the temperature of the instrument, the compensation is made by using a calibration table having a reference junction of 0°F. Compensation is made by ascertaining the millivoltage for the actual cold-junction temperature of the thermocouple and adjusting the compensation-resistance knob to this value. The actual millivoltage of the hot junction then can be read directly. These 32°F. reference-junction tables may also be used for this type of compensation by numerically adding the millivoltage at 0°F. to all of the millivoltage figures in the table.

The potentiometer has a fairly high cost and the new electronic high-speed instruments are still more expensive. The latter also requires a man skilled in electronics for care and maintenance. But where rapid, accurate temperature checks are required, such as in the open hearth bath and hot-rolled products, this high-speed unit is indispensable. Since the electronic potentiometer contains no galvanometer, this instrument can withstand considerable vibration.

Dirt is also a big headache with the mechanical instruments, particularly on the slide-wire, contacts and galvanometer. Some of the instruments are mechanically standardized, but should be checked to insure the presence of a good dry cell. When located in the vicinity of excessive dirt or corrosive atmosphere, an additional case should cover the entire controlling and recording unit. Where a large number of furnaces are to be controlled, such as in an annealing department, the instruments are mounted in dustproof station rooms where the pyrometer man may open up the instrument for inspection. This room can also be provided with storage space for control records and charts.

Automatic Control Terminology

This terminology is based on definitions adopted by the American Society of Mechanical Engineers Committee on Industrial Instruments and Regulators. The definitions refer to temperature as the variable desired to control.

AUTOMATIC CONTROLLER is an apparatus which measures the temperature in degrees and operates to maintain within limits this measured value.

CONTROLLED VARIABLE is the temperature of a furnace which is measured and controlled by the automatic controller.

CONTROLLED MEDIUM is the furnace whose temperature is to be controlled.

CONTROL AGENT is the electric power or fuel input which is varied by the automatic controller.

SET POINT is that value of temperature which it is desired to maintain.

DEVIATION is the difference at any moment between the actual furnace temperature and the set point.

CONTROL POINT is that average value of the furnace temperature which the automatic controller actually maintains under steady load conditions. (Not synonymous with set point.)

LOAD CHANGE is any fluctuation in the loading of a furnace or in the power voltage or fuel pressure which requires a readjustment of the heat input to maintain the control point.

OFFSET is the difference between the set point and the control point of a proportional controller caused by a sustained load change.

CYCLING (oscillation) is a periodic change in the furnace temperature from one value to another.

MEASURING MEANS consist of a thermocouple, and a resistance thermometer, or radiation pyrometer, which transmits the degree of furnace temperature to the controller.

CONTROLLING MEANS consists of those elements of an automatic controller which are involved in producing a corrective action.

PRIMARY CONTROL ELEMENT is that part of the controlling means by which the output of a thermocouple or other detector is converted into controlling action.

FINAL CONTROL ELEMENT is the fuel valve or electrical contactor which directly controls the fuel or power input respectively.

PROPORTIONAL CONTROL is a type of control with a corrective action which maintains a fixed and continuous relation between the position of the fuel valve and the value of the furnace temperature within a working range.

RESTORING MECHANISM is that part of an automatic controller which connects the thermocouple to the controlling means in such a manner as to reduce temperature fluctuations after a load change and to achieve stability in the shortest time.

PROPORTIONAL BAND is the number of degrees temperature that must be varied to cause the fuel valve of a furnace to move from closed to open position.

Delaying or retarding effects associated with industrial process control are caused by capacitance, resistance and dead time.

A. *Capacitance* is the change in quantity contained per unit of change in some reference variable. It may be considered the degrees temperature rise of a furnace per Btu heat input.

B. *Resistance* is opposition to flow or the inability of a material to conduct heat.

C. *Dead time* is any definite delay period between two related actions; for example, the length of time after the heat input has been turned on until the furnace thermocouple detects the change.

REFERENCES

General Works and Articles on Pyrometry

Pyrometry of Solids and Surfaces, R. B. Sosman, American Society for Metals, 1940, Cleveland.

Pyrometry, W. P. Wood and J. M. Cork, McGraw-Hill, 1941, New York.

Metals Handbook, 1939 and 1948 Editions, American Society for Metals.

"Overshooting Prevented in Temperature Control," M. J. Manjoine, *Iron Age,* May 3, 1945.

"A Photographic Investigation of the Brightness Temperatures of Liquid Steel Streams," J. A. Hall, *Journal of the Iron and Steel Institute,* January 1947, Vol. 155, Part I.

"Recent Developments in the Pyrometry of Liquid Iron and Steel," H. T. Clark, *Iron and Steel Engineer,* February 1946.

"Measurement of Heating Rates in Reheating Furnaces," J. W. Percy, *Iron and Steel Engineer,* July 1947.

Descriptive Literature from the following companies:

Bristol Co.
Brown Instrument Co.
Foxboro Co.
Claud S. Gordon Co.
Hoskins Mfg. Co.
Illinois Testing Laboratories Inc.
Leeds and Northrup Co.
Portable Products Corp.
Pyrometer Instrument Co.
Wheelco Instruments Co.

APPENDIX

Copper-Constantan Thermocouple

To Convert Degrees Fahrenheit to Millivolts

Cold Junction 32° F.

Type C.C.

Deg. Fahr.	−200	−100	−0	+0	+100	+200	+300	+400	+500	Deg. Fahr.
				MILLIVOLTS						
0	−4.11	−2.56	−0.671	−0.671	1.52	3.97	6.64	9.52	12.57	0
5	−4.18	−2.65	−0.773	−0.568	1.63	4.09	6.78	9.67	12.73	5
10	−4.25	−2.73	−0.874	−0.465	1.75	4.22	6.92	9.82	12.88	10
15	−4.31	−2.82	−0.975	−0.360	1.87	4.35	7.06	9.97	13.04	15
20	−4.38	−2.90	−1.07	−0.255	1.99	4.48	7.21	10.12	13.20	20
25	−4.43	−2.98	−1.17	−0.149	2.11	4.61	7.35	10.27	13.36	25
30	−4.50	−3.06	−1.27	−0.043	2.23	4.75	7.49	10.42	13.52	30
35	−4.57	−3.14	−1.37	0.064	2.35	4.88	7.63	10.57	13.67	35
40	−4.63	−3.22	−1.47	0.172	2.47	5.01	7.77	10.72	13.83	40
45	−4.69	−3.30	−1.56	0.281	2.59	5.15	7.92	10.87	13.99	45
50	−4.75	−3.38	−1.66	0.389	2.71	5.28	8.06	11.03	14.15	50
55	−4.80	−3.46	−1.75	0.499	2.83	5.41	8.20	11.18	14.31	55
60	−4.86	−3.53	−1.84	0.609	2.96	5.55	8.35	11.33	14.47	60
65	−4.92	−3.61	−1.94	0.720	3.08	5.68	8.49	11.49	14.63	65
70	−4.97	−3.68	−2.03	0.832	3.21	5.82	8.64	11.64	14.79	70
75	−5.03	−3.76	−2.12	0.944	3.33	5.96	8.79	11.79	14.96	75
80	−5.08	−3.83	−2.21	1.06	3.46	6.09	8.93	11.95	15.12	80
85	−5.13	−3.90	−2.30	1.17	3.58	6.23	9.08	12.10	15.28	85
90	−5.18	−3.97	−2.39	1.29	3.71	6.37	9.23	12.26	15.44	90
95	−5.23	−4.04	−2.47	1.40	3.84	6.51	9.37	12.41	15.61	95
100	−5.28	−4.11	−2.56	1.52	3.97	6.64	9.52	12.57	15.77	100
M. V. per °F.	0.0117	0.0155	0.0189	0.0219	0.0245	0.0268	0.0287	0.0305	0.0320	M. V. per °F.

Iron-Constantan Thermocouple

To Convert Degrees Fahrenheit to Millivolts

Cold Junction 32° F.

Type I.C.

Deg. Fahr.	0	100	200	300	400	500	600	700	800	900	1000	Deg. Fahr.
	MILLIVOLTS											
0	− 0.922	1.96	4.92	7.95	11.02	14.09	17.16	20.23	23.31	26.41*	29.54	0
5	− 0.778	2.11	5.07	8.10	11.18	14.24	17.31	20.38	23.46	26.57	29.70	5
10	− 0.634	2.26	5.22	8.26	11.32	14.40	17.47	20.54	23.62	26.72	29.86	10
15	− 0.490	2.40	5.37	8.41	11.48	14.55	17.62	20.69	23.77	26.88	30.02	15
20	− 0.346	2.55	5.53	8.56	11.63	14.70	17.77	20.85	23.93	27.04	30.18	20
25	− 0.202	2.70	5.68	8.72	11.79	14.86	17.93	21.00	24.08	27.19	30.34	25
30	− 0.058	2.85	5.83	8.87	11.94	15.01	18.08	21.15	24.24	27.35	30.50	30
35	0.088	3.00	5.98	9.02	12.09	15.16	18.23	21.31	24.39	27.50	30.66	35
40	0.228	3.14	6.13	9.18	12.25	15.32	18.39	21.46	24.55	27.66	30.82	40
45	0.378	3.29	6.28	9.33	12.40	15.47	18.54	21.62	24.70	27.82	30.98	45
50	0.518	3.44	6.44	9.48	12.55	15.62	18.69	21.77	24.86	27.97	31.14	50
55	0.658	3.59	6.58	9.64	12.71	15.78	18.85	21.92	25.01	28.13	31.30	55
60	0.808	3.74	6.74	9.79	12.86	15.93	19.00	22.08	25.17	28.29	31.46	60
65	0.948	3.88	6.89	9.94	13.02	16.08	19.15	22.23	25.32	28.44	31.62	65
70	1.10	4.03	7.04	10.10	13.17	16.24	19.31	22.39	25.48	28.60	31.78	70
75	1.24	4.18	7.19	10.25	13.32	16.39	19.46	22.54	25.63	28.76	31.94	75
80	1.38	4.33	7.34	10.40	13.48	16.55	19.62	22.69	25.79	28.91	32.10	80
85	1.53	4.48	7.50	10.55	13.63	16.70	19.77	22.85	25.94	29.07	32.26	85
90	1.67	4.62	7.65	10.71	13.78	16.85	19.92	23.00	26.10	29.23	32.42	90
95	1.82	4.77	7.80	10.86	13.94	17.01	20.08	23.16	26.25	29.38	32.58	95
100	1.96	4.92	7.95	11.02	14.03	17.16	20.23	23.31	26.41	29.54	32.74	100
M. V. per °F.	0.0288	0.0296	0.0303	0.0307	0.0307	0.0307	0.0307	0.0308	0.0310	0.0313	0.0320	M. V. per °F.

Deg. Fahr.	1100	1200	1300	1400	1500	1600	1700	1800	1900	2000	2100	Deg. Fahr.
	MILLIVOLTS											
0	32.74	36.04	39.46	42.96	46.48	50.00	53.52	57.04	60.56	64.08	67.60	0
5	32.90	36.21	39.64	43.14	46.66	50.18	53.70	57.22	60.74	64.26	67.78	5
10	33.07	36.38	39.81	43.31	46.83	50.35	53.87	57.39	60.91	64.43	67.95	10
15	33.23	36.55	39.98	43.49	47.01	50.53	54.05	57.57	61.09	64.61	68.13	15
20	33.40	36.72	40.16	43.66	47.18	50.70	54.22	57.74	61.26	64.78	68.30	20
25	33.56	36.89	40.34	43.84	47.36	50.88	54.40	57.92	61.44	64.96	68.48	25
30	33.73	37.07	40.51	44.02	47.54	51.06	54.58	58.10	61.62	65.14	68.66	30
35	33.89	37.24	40.68	44.19	47.71	51.23	54.75	58.27	61.79	65.31	68.83	35
40	34.06	37.41	40.86	44.37	47.89	51.41	54.93	58.45	61.97	65.49	69.01	40
45	34.22	37.58	41.03	44.54	48.06	51.58	55.10	58.62	62.14	65.66	69.18	45
50	34.39	37.75	41.21	44.72	48.24	51.76	55.28	58.80	62.32	65.84	69.36	50
55	34.55	37.92	41.38	44.90	48.42	51.94	55.46	58.98	62.50	66.02	69.54	55
60	34.72	38.09	41.56	45.07	48.59	52.11	55.63	59.15	62.67	66.19	69.71	60
65	34.88	38.26	41.73	45.25	48.77	52.29	55.81	59.33	62.85	66.37	69.89	65
70	35.05	38.43	41.91	45.42	48.94	52.46	55.98	59.50	63.02	66.54	70.06	70
75	35.21	38.60	42.08	45.60	49.12	52.64	56.16	59.68	63.20	66.72	70.24	75
80	35.38	38.78	42.26	45.78	49.30	52.82	56.34	59.86	63.38	66.90	70.42	80
85	35.54	38.95	42.43	45.95	49.47	52.99	56.51	60.03	63.55	67.07	70.59	85
90	35.71	39.12	42.61	46.13	49.65	53.17	56.69	60.21	63.73	67.25	70.77	90
95	35.87	39.29	42.78	46.30	49.82	53.34	56.86	60.38	63.90	67.42	70.94	95
100	36.04	39.46	42.96	46.48	50.00	53.52	57.04	60.56	64.08	67.60	71.12	100
M. V. per °F.	0.0330	0.0342	0.0350	0.0352	0.0352	0.0352	0.0352	0.0352	0.0352	0.0352	0.0352	M. V. per °F.

Chromel-Alumel Thermocouple
To Convert Degrees Fahrenheit to Millivolts
Cold Junction 32°F.
Type M.A.

Deg. F.	0	100	200	300	400	500	600	700	800	900	1000	1100	1200	Deg. F.
						MILLIVOLTS								
0	−0.68	1.52	3.82	6.09	8.31	10.56	12.85	15.18	17.52	19.88	22.25	24.62	26.98	0
5	−0.58	1.63	3.93	6.20	8.42	10.67	12.96	15.29	17.63	20.00	22.37	24.74	27.10	5
10	−0.47	1.74	4.05	6.31	8.53	10.79	13.08	15.41	17.75	20.12	22.49	24.85	27.21	10
15	−0.37	1.85	4.16	6.42	8.64	10.90	13.19	15.52	17.87	20.24	22.60	24.97	27.33	15
20	−0.26	1.97	4.28	6.53	8.76	11.02	13.31	15.64	17.99	20.36	22.72	25.09	27.45	20
25	−0.15	2.08	4.39	6.64	8.87	11.13	13.43	15.76	18.10	20.47	22.84	25.21	27.57	25
30	−0.04	2.20	4.51	6.75	8.98	11.25	13.55	15.88	18.22	20.59	22.96	25.33	27.68	30
35	0.07	2.31	4.62	6.86	9.09	11.36	13.67	16.00	18.34	20.71	23.08	25.45	27.80	35
40	0.18	2.43	4.74	6.98	9.20	11.47	13.78	16.11	18.46	20.83	23.20	25.57	27.92	40
45	0.29	2.54	4.85	7.09	9.31	11.58	13.89	16.23	18.58	20.95	23.32	25.69	28.04	45
50	0.40	2.66	4.97	7.20	9.43	11.70	14.01	16.35	18.70	21.07	23.43	25.80	28.15	50
55	0.51	2.77	5.08	7.31	9.55	11.81	14.12	16.47	18.81	21.18	23.55	25.92	28.27	55
60	0.62	2.89	5.19	7.42	9.66	11.93	14.24	16.58	18.93	21.30	23.67	26.04	28.39	60
65	0.73	3.00	5.30	7.53	9.77	12.04	14.36	16.70	19.05	21.42	23.79	26.16	28.51	65
70	0.84	3.12	5.42	7.64	9.88	12.16	14.48	16.82	19.17	21.54	23.91	26.27	28.62	70
75	0.95	3.24	5.53	7.75	9.99	12.27	14.60	16.93	19.29	21.66	24.02	26.39	28.74	75
80	1.06	3.36	5.64	7.87	10.11	12.39	14.71	17.05	19.41	21.78	24.14	26.51	28.86	80
85	1.17	3.48	5.75	7.98	10.22	12.50	14.83	17.17	19.52	21.89	24.26	26.63	28.98	85
90	1.29	3.59	5.87	8.09	10.33	12.62	14.94	17.29	19.64	22.01	24.38	26.74	29.09	90
95	1.40	3.70	5.98	8.20	10.44	12.73	15.06	17.40	19.76	22.13	24.50	26.86	29.21	95
100	1.52	3.82	6.09	8.31	10.56	12.85	15.18	17.52	19.88	22.25	24.62	26.98	29.33	100
M. V. per °F.	0.0220	0.0230	0.0227	0.0222	0.0225	0.0229	0.0233	0.0234	0.0236	0.0237	0.0237	0.0236	0.0235	M. V. per °F.

Deg. F.	1300	1400	1500	1600	1700	1800	1900	2000	2100	2200	2300	2400	Deg. F.
					MILLIVOLTS								
0	29.33	31.65	33.94	36.20	38.43	40.62	42.77	44.89	46.97	49.01	51.00	52.95	0
5	29.45	31.77	34.06	36.31	38.54	40.73	42.88	45.00	47.08	49.11	51.10	53.05	5
10	29.56	31.88	34.17	36.42	38.65	40.83	42.98	45.10	47.18	49.21	51.20	53.14	10
15	29.68	32.00	34.29	36.54	38.76	40.94	43.09	45.20	47.28	49.31	51.30	53.24	15
20	29.79	32.11	34.40	36.65	38.87	41.05	43.20	45.31	47.38	49.41	51.39	53.33	20
25	29.91	32.23	34.51	36.76	38.98	41.16	43.31	45.41	47.49	49.51	51.49	53.43	25
30	30.02	32.34	34.62	36.87	39.09	41.27	43.41	45.52	47.59	49.61	51.59	53.52	30
35	30.14	32.46	34.74	36.99	39.20	41.38	43.52	45.62	47.69	49.71	51.69	53.62	35
40	30.26	32.57	34.85	37.10	39.31	41.48	43.62	45.73	47.79	49.81	51.78	53.71	40
45	30.38	32.69	34.97	37.21	39.42	41.59	43.73	45.83	47.89	49.91	51.88	53.81	45
50	30.49	32.80	35.08	37.32	39.53	41.70	43.83	45.93	47.99	50.01	51.98	53.90	50
55	30.61	32.92	36.19	37.43	39.64	41.81	43.94	46.04	48.10	50.11	52.08	54.00	55
60	30.72	33.03	35.30	37.54	39.75	41.91	44.04	46.14	48.20	50.21	52.17	54.09	60
65	30.84	33.15	35.42	37.65	39.86	42.02	44.15	46.25	48.30	50.31	52.27	54.19	65
70	30.96	33.26	35.53	37.76	39.96	42.13	44.26	46.35	48.40	50.41	52.37	54.28	70
75	31.08	33.38	35.64	37.88	40.07	42.24	44.37	46.46	48.51	50.51	52.47	54.38	75
80	31.19	33.49	35.75	37.99	40.18	42.34	44.47	46.56	48.61	50.61	52.56	54.47	80
85	31.31	33.60	35.87	38.10	40.29	42.45	44.58	46.66	48.71	50.71	52.66	54.57	85
90	31.42	33.71	35.98	38.21	40.40	42.56	44.68	46.76	48.81	50.80	52.75	54.66	90
95	31.54	33.83	36.09	38.32	40.51	42.67	44.79	46.87	48.91	50.90	52.85	54.76	95
100	31.65	33.94	36.20	38.43	40.62	42.77	44.89	46.97	49.01	51.00	52.95	54.85	100
M. V. per °F.	0.0232	0.0229	0.0226	0.0223	0.0219	0.0215	0.0212	0.0208	0.0204	0.0199	0.0195	0.0190	M. V. per °F.

Appendix

Platinum-Platinum + 10% Rhodium Thermocouple
To Convert Degrees Fahrenheit of Millivolts
Cold Junction 32° F.
Type HER

Deg.F.	0	100	200	300	400	500	600	700	800	900	1000	1100	1200	1300	1400	1500	Deg.F
	MILLIVOLTS																
0	−0.0920	0.221	0.595	1.016	1.473	1.956	2.457	2.975	3.505	4.044	4.594	5.155	5.725	6.307	6.898	7.500	0
5	−0.0778	0.239	0.615	1.038	1.497	1.980	2.482	3.001	3.532	4.071	4.621	5.183	5.754	6.336	6.928	7.530	5
10	−0.0636	0.257	0.635	1.060	1.521	2.005	2.508	3.028	3.559	4.098	4.649	5.211	5.783	6.366	6.958	7.561	10
15	−0.0494	0.274	0.655	1.082	1.544	2.030	2.534	3.054	3.585	4.125	4.677	5.239	5.812	6.395	6.987	7.591	15
20	−0.0351	0.292	0.676	1.105	1.568	2.055	2.560	3.081	3.612	4.153	4.705	5.268	5.841	6.424	7.017	7.621	20
25	−0.0207	0.310	0.696	1.127	1.592	2.080	2.585	3.107	3.639	4.180	4.733	5.296	5.869	6.453	7.047	7.651	25
30	−0.0060	0.328	0.717	1.150	1.616	2.105	2.611	3.133	3.667	4.208	4.761	5.324	5.898	6.483	7.077	7.682	30
35	0.0090	0.346	0.737	1.173	1.640	2.130	2.637	3.159	3.694	4.235	4.789	5.352	5.927	6.512	7.107	7.713	35
40	0.0243	0.365	0.758	1.196	1.664	2.155	2.663	3.186	3.720	4.263	4.817	5.381	5.956	6.542	7.137	7.744	40
45	0.0398	0.383	0.779	1.218	1.688	2.180	2.689	3.212	3.747	4.290	4.845	5.409	5.985	6.571	7.167	7.774	45
50	0.0555	0.401	0.800	1.241	1.712	2.205	2.715	3.239	3.774	4.318	4.873	5.438	6.015	6.601	7.198	7.805	50
55	0.0714	0.420	0.821	1.264	1.736	2.230	2.741	3.266	3.801	4.345	4.901	5.466	6.044	6.630	7.228	7.835	55
60	0.0875	0.439	0.843	1.287	1.760	2.255	2.767	3.293	3.828	4.373	4.929	5.495	6.073	6.660	7.258	7.866	60
65	0.104	0.458	0.864	1.310	1.784	2.280	2.793	3.319	3.855	4.400	4.957	5.524	6.102	6.690	7.288	7.897	65
70	0.120	0.477	0.886	1.333	1.808	2.305	2.819	3.346	3.882	4.428	4.985	5.553	6.132	6.720	7.318	7.928	70
75	0.137	0.496	0.907	1.356	1.832	2.330	2.845	3.372	3.909	4.455	5.013	5.581	6.161	6.749	7.348	7.958	75
80	0.153	0.516	0.929	1.380	1.857	2.356	2.871	3.399	3.936	4.483	5.042	5.610	6.190	6.779	7.379	7.989	80
85	0.170	0.535	0.950	1.403	1.881	2.381	2.897	3.425	3.963	4.511	5.070	5.638	6.219	6.808	7.409	8.020	85
90	0.187	0.555	0.972	1.426	1.906	2.406	2.923	3.452	3.990	4.539	5.098	5.667	6.249	6.838	7.439	8.051	90
95	0.204	0.575	0.994	1.449	1.931	2.431	2.943	3.478	4.017	4.566	5.126	5.696	6.278	6.868	7.469	8.081	95
100	0.221	0.595	1.016	1.473	1.956	2.457	2.975	3.505	4.044	4.594	5.155	5.725	6.307	6.898	7.500	8.112	100
M. V. per °F.	0.00325	0.00374	0.00421	0.00457	0.00483	0.00501	0.00518	0.00530	0.00539	0.00550	0.00561	0.00570	0.00582	0.00591	0.00602	0.00612	M. V. per °F.

Deg.F.	1600	1700	1800	1900	2000	2100	2200	2300	2400	2500	2600	2700	2800	2900	3000	3100	Deg.F.
	MILLIVOLTS																
0	8.112	8.734	9.365	10.007	10.657	11.316	11.977	12.642	13.305	13.968	14.629	15.288	15.943	16.596	17.247	17.892	0
5	8.143	8.765	9.397	10.039	10.690	11.349	12.010	12.675	13.339	14.001	14.662	15.321	15.976	16.629	17.279	17.920	5
10	8.174	8.796	9.429	10.071	10.723	11.382	12.043	12.708	13.372	14.034	14.695	15.353	16.009	16.661	17.311	17.957	10
15	8.205	8.827	9.461	10.103	10.756	11.415	12.076	12.741	13.405	14.067	14.728	15.386	16.042	16.694	17.344	17.989	15
20	8.236	8.859	9.493	10.136	10.789	11.448	12.110	12.775	13.438	14.100	14.761	15.418	16.074	16.726	17.376	18.021	20
25	8.267	8.890	9.525	10.168	10.822	11.481	12.143	12.808	13.472	14.133	14.794	15.451	16.107	16.759	17.408		25
30	8.298	8.922	9.557	10.201	10.855	11.514	12.177	12.841	13.505	14.166	14.826	15.484	16.139	16.791	17.440		30
35	8.329	8.953	9.589	10.233	10.887	11.547	12.210	12.874	13.538	14.200	14.859	15.517	16.172	16.824	17.473		35
40	8.360	8.895	9.621	10.263	10.920	11.580	12.243	12.907	13.571	14.233	14.892	15.550	16.205	16.856	17.505		40
45	8.391	9.016	9.653	10.298	10.953	11.613	12.276	12.941	13.604	14.266	14.925	15.583	16.238	16.889	17.538		45
50	8.422	9.048	9.685	10.331	10.986	11.646	12.310	12.974	13.637	14.299	14.958	15.615	16.270	16.922	17.570		50
55	8.453	9.079	9.717	10.363	11.019	11.679	12.343	13.007	13.670	14.332	14.991	15.648	16.303	16.955	17.602		55
60	8.484	9.111	9.749	10.396	11.052	11.712	12.376	13.040	13.703	14.365	15.024	15.680	16.335	16.987	17.634		60
65	8.515	9.143	9.781	10.428	11.085	11.745	12.409	13.074	13.737	14.398	15.057	15.713	16.368	17.020	17.667		65
70	8.546	9.175	9.813	10.461	11.118	11.778	12.442	13.107	13.770	14.431	15.090	15.746	16.401	17.052	17.699		70
75	8.577	9.206	9.845	10.493	11.151	11.811	12.475	13.140	13.803	14.464	15.123	15.779	16.434	17.085	17.731		75
80	8.609	9.238	9.877	10.526	11.184	11.844	12.509	13.173	13.836	14.497	15.156	15.812	16.466	17.117	17.763		80
85	8.640	9.270	9.909	10.558	11.217	11.877	12.542	13.206	13.869	14.530	15.189	15.845	16.499	17.150	17.796		85
90	8.671	9.302	9.942	10.591	11.250	11.911	12.575	13.239	13.902	14.563	15.222	15.878	16.531	17.182	17.828		90
95	8.702	9.333	9.974	10.624	11.283	11.944	12.608	13.272	13.935	14.596	15.255	15.911	16.564	17.215	17.860		95
100	8.734	9.365	10.007	10.657	11.316	11.977	12.642	13.305	13.968	14.629	15.288	15.943	16.596	17.247	17.892		100
M. V. per °F.	0.00622	0.00631	0.00642	0.00650	0.00659	0.00661	0.00665	0.00663	0.00663	0.00661	0.00659	0.00655	0.00653	0.00651	0.00645		M. V. per °F.

Platinum-Platinum + 13% Rhodium Thermocouple
To Convert Degrees Fahrenheit to Millivolts
Cold Junction 32°F.
Type Q.R.

Deg. F.	0	100	200	300	400	500	600	700	800	900	1000	1100	1200	1300	1400	1500	Deg. F.
	MILLIVOLTS																
0	−0.0890	0.220	0.596	1.030	1.504	2.012	2.546	3.102	3.675	4.263	4.867	5.486	6.122	6.773	7.438	8.118	0
5	−0.0756	0.237	0.616	1.052	1.529	2.038	2.573	3.130	3.704	4.293	4.897	5.517	6.154	6.806	7.472	8.152	5
10	−0.0621	0.255	0.637	1.075	1.553	2.064	2.601	3.159	3.733	4.323	4.928	5.548	6.187	6.839	7.505	8.186	10
15	−0.0484	0.273	0.658	1.098	1.578	2.090	2.628	3.187	3.762	4.352	4.959	5.579	6.219	6.872	7.539	8.220	15
20	−0.0346	0.291	0.679	1.121	1.603	2.116	2.656	3.215	3.792	4.382	4.990	5.611	6.251	6.905	7.573	8.255	20
25	−0.0205	0.309	0.700	1.144	1.628	2.142	2.683	3.243	3.821	4.412	5.021	5.643	6.283	6.938	7.607	8.289	25
30	−0.0060	0.327	0.721	1.167	1.653	2.169	2.711	3.272	3.851	4.442	5.052	5.675	6.316	6.972	7.641	8.324	30
35	0.0090	0.345	0.742	1.190	1.678	2.196	2.739	3.300	3.880	4.472	5.083	5.706	6.348	7.005	7.674	8.359	35
40	0.0244	0.364	0.764	1.214	1.703	2.223	2.767	3.329	3.910	4.502	5.114	5.738	6.381	7.038	7.708	8.394	40
45	0.0399	0.382	0.785	1.237	1.728	2.249	2.794	3.357	3.939	4.532	5.144	5.770	6.413	7.071	7.742	8.428	45
50	0.0555	0.400	0.807	1.261	1.753	2.276	2.822	3.386	3.969	4.562	5.175	5.802	6.446	7.104	7.776	8.463	50
55	0.0712	0.419	0.829	1.285	1.778	2.303	2.850	3.415	3.998	4.592	5.206	5.834	6.478	7.137	7.810	8.498	55
60	0.0871	0.439	0.851	1.309	1.804	2.330	2.878	3.444	4.028	4.623	5.237	5.866	6.511	7.170	7.845	8.533	60
65	0.103	0.458	0.873	1.333	1.830	2.356	2.906	3.472	4.057	4.653	5.268	5.898	6.544	7.203	7.879	8.567	65
70	0.119	0.477	0.895	1.357	1.856	2.383	2.934	3.501	4.087	4.684	5.299	5.930	6.577	7.237	7.913	8.602	70
75	0.136	0.496	0.917	1.381	1.881	2.410	2.962	3.530	4.116	4.714	5.330	5.962	6.609	7.270	7.947	8.637	75
80	0.152	0.516	0.939	1.405	1.907	2.438	2.990	3.559	4.145	4.745	5.361	5.994	6.642	7.303	7.981	8.672	80
85	0.169	0.536	0.961	1.429	1.933	2.465	3.018	3.588	4.174	4.775	5.392	6.026	6.674	7.337	8.015	8.706	85
90	0.186	0.556	0.984	1.454	1.959	2.492	3.046	3.617	4.204	4.806	5.423	6.058	6.707	7.371	8.049	8.741	90
95	0.203	0.576	1.007	1.479	1.985	2.519	3.074	3.646	4.233	4.836	5.454	6.090	6.740	7.404	8.083	8.776	95
100	0.220	0.596	1.030	1.504	2.012	2.546	3.102	3.675	4.263	4.867	5.486	6.122	6.773	7.438	8.118	8.811	100
M. V. per °F.	0.00323	0.00376	0.00434	0.00474	0.00508	0.00534	0.00556	0.00573	0.00588	0.00604	0.00619	0.00636	0.00651	0.00665	0.00680	0.00693	M. V. per °F.

Deg. F.	1600	1700	1800	1900	2000	2100	2200	2300	2400	2500	2600	2700	2800	2900	3000	Deg. F.
	MILLIVOLTS															
0	8.811	9.518	10.237	10.970	11.720	12.478	13.242	14.010	14.777	15.543	16.309	17.073	17.833	18.588	19.342	0
5	8.845	9.553	10.273	11.007	11.758	12.516	13.280	14.048	14.815	15.581	16.347	17.111	17.871	18.626	19.380	5
10	8.880	9.589	10.310	11.045	11.796	12.554	13.318	14.086	14.853	15.619	16.385	17.149	17.909	18.664	19.417	10
15	8.915	9.624	10.346	11.082	11.833	12.592	13.356	14.124	14.892	15.657	16.424	17.187	17.947	18.702	19.455	15
20	8.951	9.660	10.383	11.119	11.871	12.631	13.395	14.163	14.930	15.695	16.462	17.225	17.985	18.739	19.492	20
25	8.986	9.695	10.419	11.156	11.909	12.669	13.433	14.201	14.968	15.734	16.501	17.264	18.023	18.777	19.530	25
30	9.021	9.734	10.456	11.194	11.947	12.707	13.472	14.240	15.006	15.772	16.539	17.302	18.060	18.815	19.567	30
35	9.056	9.767	10.492	11.231	11.985	12.745	13.510	14.278	15.045	15.811	16.577	17.340	18.098	18.853	19.605	35
40	9.092	9.803	10.529	11.269	12.023	12.783	13.549	14.316	15.083	15.849	16.615	17.377	18.135	18.891	19.642	40
45	9.127	9.839	10.565	11.306	12.061	12.821	13.587	14.355	15.122	15.887	16.654	17.415	18.173	18.929	19.680	45
50	9.162	9.875	10.602	11.344	12.099	12.860	13.625	14.393	15.160	15.925	16.692	17.453	18.211	18.966	19.717	50
55	9.197	9.911	10.638	11.382	12.136	12.898	13.663	14.432	15.199	15.964	16.731	17.432	18.249	19.004	19.755	55
60	9.233	9.947	10.675	11.420	12.174	12.936	13.702	14.470	15.237	16.002	16.769	17.530	18.286	19.041	19.793	60
65	9.268	9.983	10.712	11.457	12.212	12.974	13.740	14.509	15.275	16.041	16.807	17.568	18.324	19.079	19.830	65
70	9.304	10.019	10.749	11.495	12.250	13.012	13.779	14.547	15.313	16.079	16.845	17.605	18.362	19.117	19.867	70
75	9.339	10.055	10.785	11.532	12.288	13.050	13.817	14.536	15.351	16.117	16.883	17.643	18.400	19.155	19.905	75
80	9.375	10.091	10.822	11.570	12.326	13.089	13.856	14.624	15.389	16.155	16.921	17.682	18.437	19.192	19.942	80
85	9.410	10.127	10.859	11.607	12.364	13.127	13.894	14.663	15.428	16.194	16.959	17.720	18.475	19.230	19.980	85
90	9.446	10.164	10.896	11.645	12.402	13.166	13.933	14.701	15.466	16.232	16.997	17.758	18.513	19.267	20.017	90
95	9.482	10.200	10.933	11.682	12.440	13.204	13.971	14.739	15.505	16.271	17.035	17.796	18.551	19.305	20.055	95
100	9.518	10.237	10.970	11.720	12.478	13.242	14.010	14.777	15.543	16.309	17.073	17.833	18.588	19.342	20.093	100
M. V. per °F.	0.00707	0.00719	0.00733	0.00750	0.00758	0.00764	0.00768	0.00767	0.00766	0.00766	0.00764	0.00760	0.00755	0.00754	0.00751	M. V. per °F.

Conversion of Centigrade and Fahrenheit Scales

For temperatures above 0° C.

Temp. °C.	0	1	2	3	4	5	6	7	8	9
0	32.0	33.8	35.6	37.4	39.2	41.0	42.8	44.6	46.4	48.2
10	50.0	51.8	53.6	55.4	57.2	59.0	60.8	62.6	64.4	66.2
20	68.0	69.8	71.6	73.4	75.2	77.0	78.8	80.6	82.4	84.2
30	86.0	87.8	89.6	91.4	93.2	95.0	96.8	98.6	100.4	102.2
40	104.0	105.8	107.6	109.4	111.2	113.0	114.8	116.6	118.4	120.2
50	122.0	123.8	125.6	127.4	129.2	131.0	132.8	134.6	136.4	138.2
60	140.0	141.8	143.6	145.4	147.2	149.0	150.8	152.6	154.4	156.2
70	158.0	159.8	161.6	163.4	165.2	167.0	168.8	170.6	172.4	174.2
80	176.0	177.8	179.6	181.4	183.2	185.0	186.8	188.6	190.4	192.2
90	194.0	195.8	197.6	199.4	201.2	203.0	204.8	206.6	208.4	210.2
100	212.0	213.8	215.6	217.4	219.2	221.0	222.8	224.6	226.4	228.2
110	230.0	231.8	233.6	235.4	237.2	239.0	240.8	242.6	244.4	246.2
120	248.0	249.8	251.6	253.4	255.2	257.0	258.8	260.6	262.4	264.2
130	266.0	267.8	269.6	271.4	273.2	275.0	276.8	278.6	280.4	282.2
140	284.0	285.8	287.6	289.4	291.2	293.0	294.8	296.6	298.4	300.2
150	302.0	303.8	305.6	307.4	309.2	311.0	312.8	314.6	316.4	318.2
160	320.0	321.8	323.6	325.4	327.2	329.0	330.8	332.6	334.4	336.2
170	338.0	339.8	341.6	343.4	345.2	347.0	348.8	350.6	352.4	354.2
180	356.0	357.8	359.6	361.4	363.2	365.0	366.8	368.6	370.4	372.2
190	374.0	375.8	377.6	379.4	381.2	383.0	384.8	386.6	388.4	390.2
200	392.0	393.8	395.6	397.4	399.2	401.0	402.8	404.6	406.4	408.2
210	410.0	411.8	413.6	415.4	417.2	419.0	420.8	422.6	424.4	426.2
220	428.0	429.8	431.6	433.4	435.2	437.0	438.8	440.6	442.4	444.2
230	446.0	447.8	449.6	451.4	453.2	455.0	456.8	458.6	460.4	462.2
240	464.0	465.8	467.6	469.4	471.2	473.0	474.8	476.6	478.4	480.2
250	482.0	483.8	485.6	487.4	489.2	491.0	492.8	494.6	496.4	498.2
260	500.0	501.8	503.6	505.4	507.2	509.0	510.8	512.6	514.4	516.2
270	518.0	519.8	521.6	523.4	525.2	527.0	528.8	530.6	532.4	534.2
280	536.0	537.8	539.6	541.4	543.2	545.0	546.8	548.6	550.4	552.2
290	554.0	555.8	557.6	559.4	561.2	563.0	564.8	566.6	568.4	570.2
300	572.0	573.8	575.6	577.4	579.2	581.0	582.8	584.6	586.4	588.2
310	590.0	591.8	593.6	595.4	597.2	599.0	600.8	602.6	604.4	606.2
320	608.0	609.8	611.6	613.4	615.2	617.0	618.8	620.6	622.4	624.2
330	626.0	627.8	629.6	631.4	633.2	635.0	636.8	638.6	640.4	642.2
340	644.0	645.8	647.6	649.4	651.2	653.0	654.8	656.6	658.4	660.2
350	662.0	663.8	665.6	667.4	669.2	671.0	672.8	674.6	676.4	678.2
360	680.0	681.8	683.6	685.4	687.2	689.0	690.8	692.6	694.4	696.2
370	698.0	699.8	701.6	703.4	705.2	707.0	708.8	710.6	712.4	714.2
380	716.0	717.8	719.6	721.4	723.2	725.0	726.8	728.6	730.4	732.2
390	734.0	735.8	737.6	739.4	741.2	743.0	744.8	746.6	748.4	750.2
400	752.0	753.8	755.6	757.4	759.2	761.0	762.8	764.6	766.4	768.2
410	770.0	771.8	773.6	775.4	777.2	779.0	780.8	782.6	784.4	786.2
420	788.0	789.8	791.6	793.4	795.2	797.0	798.8	800.6	802.4	804.2
430	806.0	807.8	809.6	811.4	813.2	815.0	816.8	818.6	820.4	822.2
440	824.0	825.8	827.6	829.4	831.2	833.0	834.8	836.6	838.4	840.2

Conversion of Centigrade and Fahrenheit Scales
(Continued)

450	842.0	843.8	845.6	847.4	849.2	851.0	852.8	854.6	856.4	858.2
460	860.0	861.8	863.6	865.4	867.2	869.0	870.8	872.6	874.4	876.2
470	878.0	879.8	881.6	883.4	885.2	887.0	888.8	890.6	892.4	894.2
480	896.0	897.8	899.6	901.4	903.2	905.0	906.8	908.6	910.4	912.2
490	914.0	915.8	917.6	919.4	921.2	923.0	924.8	926.6	928.4	930.2
500	932.0	933.8	935.6	937.4	939.2	941.0	942.8	944.6	946.4	948.2
510	950.0	951.8	953.6	955.4	957.2	959.0	960.8	962.6	964.4	966.2
520	968.0	969.8	971.6	973.4	975.2	977.0	978.8	980.6	982.4	984.2
530	986.0	987.8	989.6	991.4	993.2	995.0	996.8	998.6	1000.4	1002.2
540	1004.0	1005.8	1007.6	1009.4	1011.2	1013.0	1014.8	1016.6	1018.4	1020.2
550	1022.0	1023.8	1025.6	1027.4	1029.2	1031.0	1032.8	1034.6	1036.4	1038.2
560	1040.0	1041.8	1043.6	1045.4	1047.2	1049.0	1050.8	1052.6	1054.4	1056.2
570	1058.0	1059.8	1061.6	1063.4	1065.2	1067.0	1068.8	1070.6	1072.4	1074.2
580	1076.0	1077.8	1079.6	1081.4	1083.2	1085.0	1086.8	1088.6	1090.4	1092.2
590	1094.0	1095.8	1097.6	1099.4	1101.2	1103.0	1104.8	1106.6	1108.4	1110.2
600	1112.0	1113.8	1115.6	1117.4	1119.2	1121.0	1122.8	1124.6	1126.4	1128.2
610	1130.0	1131.8	1133.6	1135.4	1137.2	1139.0	1140.8	1142.6	1144.4	1146.2
620	1148.0	1149.8	1151.6	1153.4	1155.2	1157.0	1158.8	1160.6	1162.4	1164.2
630	1166.0	1167.8	1169.6	1171.4	1173.2	1175.0	1176.8	1178.6	1180.4	1182.2
640	1184.0	1185.8	1187.6	1189.4	1191.2	1193.0	1194.8	1196.6	1198.4	1200.2
650	1202.0	1203.8	1205.6	1207.4	1209.2	1211.0	1212.8	1214.6	1216.4	1218.2
660	1220.0	1221.8	1223.6	1225.4	1227.2	1229.0	1230.8	1232.6	1234.4	1236.2
670	1238.0	1239.8	1241.6	1243.4	1245.2	1247.0	1248.8	1250.6	1252.4	1254.2
680	1256.0	1257.8	1259.6	1261.4	1263.2	1265.0	1266.8	1268.6	1270.4	1272.2
690	1274.0	1275.8	1277.6	1279.4	1281.2	1283.0	1284.8	1286.6	1288.4	1290.2
700	1292.0	1293.8	1295.6	1297.4	1299.2	1301.0	1302.8	1304.6	1306.4	1308.2
710	1310.0	1311.8	1313.6	1315.4	1317.2	1319.0	1320.8	1322.6	1324.4	1326.2
720	1328.0	1329.8	1331.6	1333.4	1335.2	1337.0	1338.8	1340.6	1342.4	1344.2
730	1346.0	1347.8	1349.6	1351.4	1353.2	1355.0	1356.8	1358.6	1360.4	1362.2
740	1364.0	1365.8	1367.6	1369.4	1371.2	1373.0	1374.8	1376.6	1378.4	1380.2
750	1382.0	1383.8	1385.6	1387.4	1389.2	1391.0	1392.8	1394.6	1396.4	1398.2
760	1400.0	1401.8	1403.6	1405.4	1407.2	1409.0	1410.8	1412.6	1414.4	1416.2
770	1418.0	1419.8	1421.6	1423.4	1426.2	1427.0	1428.8	1430.6	1432.4	1434.2
780	1436.0	1437.8	1439.6	1441.4	1443.2	1445.0	1446.8	1448.6	1450.4	1452.2
790	1454.0	1455.8	1457.6	1459.4	1461.2	1463.0	1464.8	1466.6	1468.4	1470.2
800	1472.0	1473.8	1475.6	1477.4	1479.2	1481.0	1482.8	1484.6	1486.4	1488.2
810	1490.0	1491.8	1493.6	1495.4	1497.2	1499.0	1500.8	1502.6	1504.4	1506.2
820	1508.0	1509.9	1511.6	1513.4	1515.2	1517.0	1518.8	1520.6	1522.4	1524.2
830	1526.0	1527.8	1529.6	1531.4	1533.2	1535.0	1536.8	1538.6	1540.4	1542.2
840	1544.0	1545.8	1547.6	1549.4	1551.2	1553.0	1554.8	1556.6	1558.4	1560.2
850	1562.0	1563.8	1565.6	1567.4	1569.2	1571.0	1572.8	1574.6	1576.4	1578.2
860	1580.0	1581.8	1583.6	1585.4	1587.2	1589.0	1590.8	1592.6	1594.4	1596.2
870	1598.0	1599.8	1601.6	1603.4	1605.2	1607.0	1608.8	1610.6	1612.4	1614.2
880	1616.0	1617.8	1619.6	1621.4	1623.2	1625.0	1626.8	1628.6	1630.4	1632.2
890	1634.0	1635.8	1637.6	1639.4	1641.2	1643.0	1644.8	1646.6	1648.4	1650.2

Conversion of Centigrade and Fahrenheit Scales
(Continued)

For temperatures above 0° C.

Temp. °C.	0	1	2	3	4	5	6	7	8	9
900	1652.0	1653.8	1655.6	1657.4	1659.2	1661.0	1662.8	1664.6	1666.4	1668.2
910	1670.0	1671.8	1673.6	1675.4	1677.2	1679.0	1680.8	1682.6	1684.4	1686.2
920	1688.0	1689.8	1691.6	1693.4	1695.2	1697.0	1698.8	1700.6	1702.4	1704.2
930	1706.0	1707.8	1709.6	1711.4	1713.2	1715.0	1716.8	1718.6	1720.4	1722.2
940	1724.0	1725.8	1727.6	1729.4	1731.2	1733.0	1734.8	1736.6	1738.4	1740.2
950	1742.0	1743.8	1745.6	1747.4	1749.2	1751.0	1752.8	1754.6	1756.4	1758.2
960	1760.0	1761.8	1763.6	1765.4	1767.2	1769.0	1770.8	1772.6	1774.4	1776.2
970	1778.0	1779.8	1781.6	1783.4	1785.2	1787.0	1788.8	1790.6	1792.4	1794.2
980	1796.0	1797.8	1799.6	1801.4	1803.2	1805.0	1806.8	1808.6	1810.4	1812.2
990	1814.0	1815.8	1817.6	1819.4	1821.2	1823.0	1824.8	1826.6	1828.4	1830.2
1000	1832.0	1833.8	1835.6	1837.4	1839.2	1841.0	1842.8	1844.6	1846.4	1848.2
1010	1850.0	1851.8	1853.6	1855.4	1857.2	1859.0	1860.8	1862.6	1864.4	1866.2
1020	1868.0	1869.8	1871.6	1873.4	1875.2	1877.0	1878.8	1880.6	1882.4	1884.2
1030	1886.0	1887.8	1889.6	1891.4	1893.2	1895.0	1896.8	1898.6	1900.4	1902.2
1040	1904.0	1905.8	1907.6	1909.4	1911.2	1913.0	1914.8	1916.6	1918.4	1920.2
1050	1922.0	1923.8	1925.6	1927.4	1929.2	1931.0	1932.8	1934.6	1936.4	1938.2
1060	1940.0	1941.8	1943.6	1945.4	1947.2	1949.0	1950.8	1952.6	1954.4	1956.2
1070	1958.0	1959.8	1961.6	1963.4	1965.2	1967.0	1968.8	1970.6	1972.4	1974.2
1080	1976.0	1977.8	1979.6	1981.4	1983.2	1985.0	1986.8	1988.6	1990.4	1992.2
1090	1994.0	1995.8	1997.6	1999.4	2001.2	2003.0	2004.8	2006.6	2008.4	2010.2
1100	2012.0	2013.8	2015.6	2017.4	2019.2	2021.0	2022.8	2024.6	2026.4	2028.2
1110	2030.0	2031.8	2033.6	2035.4	2037.2	2039.0	2040.8	2042.6	2044.4	2046.2
1120	2048.0	2049.8	2051.6	2053.4	2055.2	2057.0	2058.8	2060.6	2062.4	2064.2
1130	2066.0	2067.8	2069.6	2071.4	2073.2	2075.0	2076.8	2078.6	2080.4	2082.2
1140	2084.0	2085.8	2087.6	2089.4	2091.2	2093.0	2094.8	2096.6	2098.4	2100.2
1150	2102.0	2103.8	2105.6	2107.4	2109.2	2111.0	2112.8	2114.6	2116.4	2118.2
1160	2120.0	2121.8	2123.6	2125.4	2127.2	2129.0	2130.8	2132.6	2134.4	2136.2
1170	2138.0	2139.8	2141.6	2143.4	2145.2	2147.0	2148.8	2150.6	2152.4	2154.2
1180	2156.0	2157.8	2159.6	2161.4	2163.2	2165.0	2166.8	2168.6	2170.4	2172.2
1190	2174.0	2175.8	2177.6	2179.4	2181.2	2183.0	2184.8	2186.6	2188.4	2190.2
1200	2192.0	2193.8	2195.6	2197.4	2199.2	2201.0	2202.8	2204.6	2206.4	2208.2
1210	2210.0	2211.8	2213.6	2215.4	2217.2	2219.0	2220.8	2222.6	2224.4	2226.2
1220	2228.0	2229.8	2231.6	2233.4	2235.2	2237.0	2238.8	2240.6	2242.4	2244.2
1230	2246.0	2247.8	2249.6	2251.4	2253.2	2255.0	2256.8	2258.6	2260.4	2262.2
1240	2264.0	2265.8	2267.6	2269.4	2271.2	2273.0	2274.8	2276.6	2278.4	2280.2
1250	2282.0	2283.8	2285.6	2287.4	2289.2	2291.0	2292.8	2294.6	2296.4	2298.2
1260	2300.0	2301.8	2303.6	2305.4	2307.2	2309.0	2310.8	2312.6	2314.4	2316.2
1270	2318.0	2319.8	2321.6	2323.4	2325.2	2327.0	2328.8	2330.6	2332.4	2334.2
1280	2336.0	2337.8	2339.6	2341.4	2343.2	2345.0	2346.8	2348.6	2350.4	2352.2
1290	2354.0	2355.8	2357.6	2359.4	2361.2	2363.0	2364.8	2366.6	2368.4	2370.2
1300	2372.0	2373.8	2375.6	2377.4	2379.2	2381.0	2382.8	2384.6	2386.4	2388.2
1310	2390.0	2391.8	2393.6	2395.4	2397.2	2399.0	2400.8	2402.6	2404.4	2406.2
1320	2408.0	2409.8	2411.6	2413.4	2415.2	2417.0	2418.8	2420.6	2422.4	2424.2
1330	2426.0	2427.8	2429.6	2431.4	2433.2	2435.0	2436.8	2438.6	2440.4	2442.2
1340	2444.0	2445.8	2447.6	2449.4	2451.2	2453.0	2454.8	2456.6	2458.4	2460.2

Conversion of Centigrade and Fahrenheit Scales
(*Continued*)

1350	2462.0	2463.8	2465.6	2467.4	2469.2	2471.0	2472.8	2474.6	2476.4	2478.2
1360	2480.0	2481.8	2483.6	2485.4	2487.2	2489.0	2490.8	2492.6	2494.4	2496.2
1370	2498.0	2499.8	2501.6	2503.4	2505.2	2507.0	2508.8	2510.6	2512.4	2514.2
1380	2516.0	2517.8	2519.6	2521.4	2523.2	2525.0	2526.8	2528.6	2530.4	2532.2
1390	2534.0	2535.8	2537.6	2539.4	2541.2	2543.0	2544.8	2546.6	2548.4	2550.2
1400	2552.0	2553.8	2555.6	2557.4	2559.2	2561.0	2562.8	2564.6	2566.4	2568.2
1410	2570.0	2571.8	2573.6	2575.4	2577.2	2579.0	2580.8	2582.6	2584.4	2586.2
1420	2588.0	2589.8	2591.6	2593.4	2595.2	2597.0	2598.8	2600.6	2602.4	2604.2
1430	2606.0	2607.8	2609.6	2611.4	2613.2	2615.0	2616.8	2618.6	2620.4	2622.2
1440	2624.0	2625.8	2627.6	2629.4	2631.2	2633.0	2634.8	2636.6	2638.4	2640.2
1450	2642.0	2643.8	2645.6	2647.4	2649.2	2651.0	2652.8	2654.6	2656.4	2658.2
1460	2660.0	2661.8	2663.6	2665.4	2667.2	2669.0	2670.8	2672.6	2674.4	2676.2
1470	2678.0	2679.8	2681.6	2683.4	2685.2	2687.0	2688.8	2690.6	2692.4	2694.2
1480	2696.0	2697.8	2699.6	2701.4	2703.2	2705.0	2706.8	2708.6	2710.4	2712.2
1490	2714.0	2715.8	2717.6	2719.4	2721.2	2723.0	2724.8	2726.6	2728.4	2730.2
1500	2732.0	2733.8	2735.6	2737.4	2739.2	2741.0	2742.8	2744.6	2746.4	2748.2
1510	2750.0	2751.8	2753.6	2755.4	2757.2	2759.0	2760.8	2762.6	2764.4	2766.2
1520	2768.0	2769.8	2771.6	2773.4	2775.2	2777.0	2778.8	2780.6	2782.4	2784.2
1530	2786.0	2787.8	2789.6	2791.4	2793.2	2795.0	2796.8	2798.6	2800.4	2802.2
1540	2804.0	2805.8	2807.6	2809.4	2811.2	2813.0	2814.8	2816.6	2818.4	2820.2
1550	2822.0	2823.8	2825.6	2827.4	2829.2	2831.0	2832.8	2834.6	2836.4	2838.2
1560	2840.0	2841.8	2843.6	2845.4	2847.2	2849.0	2850.8	2852.6	2854.4	2856.2
1570	2858.0	2859.8	2861.6	2863.4	2865.2	2867.0	2868.8	2870.6	2872.4	2874.2
1580	2876.0	2877.8	2879.6	2881.4	2883.2	2885.0	2886.8	2888.6	2890.4	2892.2
1590	2894.0	2895.8	2897.6	2899.4	2901.2	2903.0	2904.8	2906.6	2908.4	2910.2
1600	2912.0	2913.8	2915.6	2917.4	2919.2	2921.0	2922.8	2924.6	2926.4	2928.2
1610	2930.0	2931.8	2933.6	2935.4	2937.2	2939.0	2940.8	2942.6	2944.4	2946.2
1620	2948.0	2949.8	2951.6	2953.4	2955.2	2957.0	2958.8	2960.6	2962.4	2964.2
1630	2966.0	2967.8	2969.6	2971.4	2973.2	2975.0	2976.8	2978.6	2980.4	2982.2
1640	2984.0	2985.8	2987.6	2989.4	2991.2	2993.0	2994.8	2996.6	2998.4	3000.2
1650	3002.0	3003.8	3005.6	3007.4	3009.2	3011.0	3012.8	3014.6	3016.4	3018.2
1660	3020.0	3021.8	3023.6	3025.4	3027.2	3029.0	3030.8	3032.6	3034.4	3036.2
1670	3038.0	3039.8	3041.6	3043.4	3045.2	3047.0	3048.8	3050.6	3052.4	3054.2
1680	3056.0	3057.8	3059.6	3061.4	3063.2	3065.0	3066.8	3068.6	3070.4	3072.2
1690	3074.0	3075.8	3077.6	3079.4	3081.2	3083.0	3084.8	3086.6	3088.4	3090.2
1700	3092.0	3093.8	3095.6	3097.4	3099.2	3101.0	3102.8	3104.6	3106.4	3108.2
1710	3110.0	3111.8	3113.6	3115.4	3117.2	3119.0	3120.8	3122.6	3124.4	3126.2
1720	3128.0	3129.8	3131.6	3133.4	3135.2	3137.0	3138.8	3140.6	3142.4	3144.2
1730	3146.0	3147.8	3149.6	3151.4	3153.2	3155.0	3156.8	3158.6	3160.4	3162.2
1740	3164.0	3165.8	3167.6	3169.4	3171.2	3173.0	3174.8	3176.6	3178.4	3180.2
1750	3182.0	3183.8	3185.6	3187.4	3189.2	3191.0	3192.8	3194.6	3196.4	3198.2
1760	3200.0	3201.8	3203.6	3205.4	3207.2	3209.0	3210.8	3212.6	3214.4	3216.2
1770	3218.0	3219.8	3221.6	3223.4	3225.2	3227.0	3228.8	3230.6	3232.4	3234.2
1780	3236.0	3237.8	3239.6	3241.4	3243.2	3245.0	3246.8	3248.6	3250.4	3252.2
1790	3254.0	3255.8	3257.6	3259.4	3261.2	3263.0	3264.8	3266.6	3268.4	3270.2

INDEX

149